T0132882

THE FLYING MATHEMATICIANS OF WORLD WAR I

The Flying Mathematicians of World War I

TONY ROYLE

McGill-Queen's University Press
Montreal & Kingston • London • Chicago

ISBN 978-0-2280-0373-1 (cloth)
ISBN 978-0-2280-0510-0 (ePDF)

Legal deposit third quarter 2020
Bibliothèque nationale du Québec

Printed in Canada on acid-free paper that is 100% ancient forest free
(100% post-consumer recycled), processed chlorine free

Library and Archives Canada Cataloguing in Publication

Title: The flying mathematicians of World War I / Tony Royle.

Names: Royle, Tony, 1959- author.

Description: Includes bibliographical references and index.

Identifiers: Canadiana (print) 2020025071X | Canadiana (ebook) 20200251090 | ISBN 9780228003731 (cloth) | ISBN 9780228005100 (ePDF)

Subjects: LCSH: Aeronautics—Great Britain—History—20th century. | LCSH: Aeronautics—Great Britain—Mathematics—History—20th century. | LCSH: Mathematicians—Great Britain—Biography. | LCSH: Air pilots—Great Britain—Biography. | LCSH: Scientists—Great Britain—Biography. | LCSH: World War, 1914-1918—Great Britain.

Classification: LCC TL526.G7 R69 2020 | DDC 629.130094109/041—dc23

This book was typeset in 10/13 LucidaBright.

"By the man in the street, mathematics is conceived of as a dull, dry subject: few laymen, possibly few mathematicians, ever realise the romance, the beauty, and the mystery of the subject."

David Hume Pinsent, 1914

Contents

Preface

Mathematics and aviation in many ways define who I am. Throughout my entire adult life one or other of these themes has been to the fore, allowing me to both make a living and pursue various passions. Half a century ago I was gripped and inspired by our attempts to reach out into space, and I imagined perhaps one day being part of such endeavour. Over time I fortuitously found myself in a position that gave me a glimmer of hope that I might be selected for a role in the space programme: I had a technical degree, a rudimentary understanding of the Russian language, and a military pilot's licence.

Of course the history books reveal that I never made it into space, but my attempts to get there opened many doors of opportunity and placed me in numerous aircraft cockpits: environments that sparked my love of flying and a desire to understand more of the engineering and mathematics underpinning aeronautics, from the fluid dynamics of Daniel Bernoulli and Leonhard Euler in the 1700s to the rocket equation that Konstantin Tsiolkovsky defined at the beginning of the twentieth century. My life as an officer and pilot in the British Royal Air Force revolved around the Lockheed C-130 Hercules transport aircraft. Its versatility meant I was deployed in a variety of roles and theatres across the globe, giving me a career that exposed me to a wide-ranging array of aviation experiences during my 16 years of service. Some missions were flown with a humanitarian focus – air-dropping or air-landing essential supplies or aid workers into areas hit by natural disaster – while others were conducted in a very different context: that of human conflict.

Much of my flying, both operational and training, was spent close to the ground: an unforgiving environment where a moment's lapse in concentration can result in disaster. I also encountered my fair share of in-flight emergencies – everything from engine fires and fuel leaks to unnerving incidents concerning extreme weather phenomena. The Hercules could land on surfaces as disparate as grass and compacted snow, operate in extremes of temperatures, refuel in mid-air if required, and deliver a variety of personnel and stores from its freight bay. Keeping on top of all the various procedures required to safely execute sorties that demanded such diversity of application was challenging but incredibly rewarding.

In stark contrast, the second half of my flying career centred around operating the large, four-engined Airbus A340 and, more latterly, the smaller twin-jet A330 aircraft for Virgin Atlantic Airways in its commercial long-haul role. This was flying of a very different nature, most of the time spent above 30,000 feet, with passenger comfort and satisfaction slightly higher up the list of operational priorities! One thing common to both phases of my piloting career, however, was the nature of those with whom I shared my experiences in the cockpit: consummate professionals, many of whom became dear friends.

In parallel with my varied and fulfilling flying career, I was fortunate to be able to further develop and explore my academic interests, studying a whole range of technical subjects. The knowledge I gleaned brought with it better insight and a much deeper understanding of aerodynamics and the forces at play when an aircraft is in flight. I always had it in mind to round off my academic studies with some sort of research, so when an opportunity arose to tackle a doctorate at the Open University, I eagerly took it; here was a chance to combine my experience in aviation with my background in mathematics and engineering and my growing interest in the historical aspects of both of these subjects. When I began to explore the genesis of British aviation and discovered the stories of the flying mathematicians, I felt humbled and a little sad: it was extraordinary what these individuals had been able to achieve, and they were clearly incredibly influential in establishing Britain's place at the global aeronautical table, yet most of their names had remained hidden from me for my entire life. Ensuring that these academic aviators had their contributions better recognized in the historical record became an important aspiration and a key motivator for my work.

The story of humankind's pursuit of flight has, of course, been told in numerous ways and by a whole host of different authors, each with a specific bias or aspiration. To add anything unique to this diverse canon of aeronautical literature, I felt I had to identify a period and an aspect of the story that would most suit my particular strengths. The obvious era to consider was that during which the general mathematics and construction techniques that had informed early attempts to aviate made the transition into what we now call aeronautical engineering. This coincided with the production and inaugural flights of the first fixed-wing, powered aircraft – machines that demanded skilled aviators at their controls.

The salient questions in my mind concerned the mathematicians and engineers responsible for this revolution. How did they become part of the story, what were their contributions, how did they interact, what were their backgrounds, what motivated them, and what became of them? Only after some preliminary investigation did I discover that some of the academics involved actually became pilots or observers themselves in order to conduct research, and this was clearly a revelation that resonated with me. I knew that as a mathematician and pilot myself, I could perhaps make more sense of what these pioneers of research in aviation must have endured, particularly as the pressures and traumas associated with World War I intensified, and specifically in relation to their cockpit experiences.

My PhD thesis was completed in 2019, and the formal requirements of such a piece of writing dictated a framing that did justice to various aspects of academic rigour. This in many ways sanitized the account, precluding the expression of personal feelings or any sense of genuine empathy with the characters at the heart of the narrative. It was out of these constraints that the idea for this book was born. I felt it would give me an opportunity to take the basic elements of the thesis as a core and integrate in a less formal but more intimate way some of my insight and experiences as a pilot – events and emotions that I could directly relate to the activities and achievements of those men and women at the coalface of aeronautical research during the first two decades of the twentieth century. This book thus sits in its own niche: a fusion of history and mathematics with personal experience and intuition.

It would be remiss and disingenuous of me to fail to mention at this juncture the works by other authors that most informed and influenced my PhD research and, therefore, this offering. While I was fortunate to discover and draw upon much unique material, neither the thesis nor this book would have been possible without reference to these extant resources. In his *Masters of Theory: Cambridge and the Rise of Mathematical Physics* (2003), historian of science Andrew Warwick provides a useful summary of how academia in Britain worked during the early twentieth century, with a focus on the function and role of Cambridge University, which supplied the bulk of the academic aeronautical influence that dominated early research and development (R&D). One gets a feel, here, for the mathematics being taught and examined in the cauldron of academic excellence that was Cambridge, and we are made fully aware of the important role of the tutors and the implications of becoming a 'Wrangler', particularly the 'Senior Wrangler'. At Cambridge University, Wrangler was the name given to a student placed in the first class of the Mathematical Tripos. The Senior Wrangler was the student placed top of the Wrangler list based on his (women were excluded from the list) academic performance, and the title brought with it considerable esteem and respect.

Broader educational context is drawn from sources such as *The History of Imperial College* (Gay 2007), *University College: London* (Bellot 1929), *The University of London and the World of Learning, 1836–1986* (Thompson 1990), and *South Kensington to Robbins* (Argles 1964), the last of which explores developments in education from the founding of the Royal College of Science in 1851 up to the Committee on Higher Education's report of 1963. In a more generic sense, the annual reports of the British Association for the Advancement of Science paint a picture of the scientific landscape behind progress in aeronautics, its meetings providing a recognized platform for influential players to make themselves heard and proffer new ideas. *The Rise of Scientific Engineering in Britain* (1985) by history of technology professor Robert A. Buchanan gives an account of how science and engineering evolved in Britain through the 1800s.

The book offering perhaps the best overall summary of events and issues in aeronautics during the early 1900s, however, is *Enigma of the Aerofoil* (2011) by the British sociologist David Bloor. While Bloor's focus is on describing the nature and derivation of the rival theories of aerodynamic lift that existed at the start of the twentieth century, the author gives us much more than this to contemplate. An inevitable consequence of his main thrust is that he is obliged to discuss and consider developments in countries other than Britain, particularly Germany, something that has been deliberately sidestepped in this book in order to concentrate specifically on aspects of the story that are pertinent to Britain.

The role that women played in all of this is a facet of history that has been relatively unexplored. It was extremely tough for any woman to gain a foothold in mathematical academia or to make any impact in the practical world of engineering in the early 1900s; nevertheless, a number would overcome the significant obstacles put in their way and make notable contributions in aeronautics. Historian Claire Jones, via her piece 'Femininity and mathematics at Cambridge circa 1900' (2010), paints a picture of the academic challenges they faced, while my own research into the workings of the technical department at the British Admiralty offers insight into their front-line engineering achievements (Royle 2017). Highlighting the important contribution that woman mathematicians made to the structural integrity of many of our early military aircraft designs provides an opportunity to recognize remarkable technical achievement in the face of adversity in many guises. If one looks carefully, their efforts can be found documented in the literature, but their stories have never been told in a coherent and concerted way – an oversight that is addressed to some extent in this book.

Sources offering a more eclectic view of the period abound, but the edited collection *The War of Guns and Mathematics* (Aubin and Goldstein 2014) stands out in discussing the international aspects of the topic, with history of mathematics professor June Barrow-Green's contribution, 'Cambridge mathematicians' responses to the First World War', being particularly illuminating. In *England and the Aeroplane* (1991), historian David Edgerton begins his account of aviation in England as far back as George Cayley in the early nineteenth century, thereby giving a more comprehensive appreciation of the history behind the developments that would eventually result in the construction of the initial tranche of World War I aircraft. This version of events is augmented by military historian Peter Reese's similar narrative: *The Men Who Gave Us Wings* (2014). *Mathematics in Victorian Britain* (Flood, Rice, and Wilson 2011) is another classic reference, and reading it makes one appreciate, among other things, the disparity between the universities of Oxford and Cambridge when considering the comparative status of mathematics: it was only really after World War I and the arrival of G.H. Hardy at Oxford in 1920 that there was any impetus towards comparability. The dominance of the Cambridge mathematicians pervades every aspect of this story, and the importance of their contributions cannot be overstated.

'The formative years of the British aircraft industry' (1969), an article by modern economic and social historian Peter Fearon, is where one can find a more quantitative look at the aviation industry during the war years: the companies involved, what they were producing in what quantities, and the associated economic data. This is reinforced by Hugh Driver's *The Birth of Military Aviation* (1997). Historian and mathematician Ben Marsden discusses the general engineering scene in Scotland in *Engineering Science in Glasgow* (1992). There are important interrelationships between aeronautical developments in the Army and Royal Navy, and maritime historian Tim Benbow's *British Naval Aviation: The First 100 Years* (2011) provides some enlightening detail.

Empathy with the mathematics of the time was established by drawing on the contemporary mathematical literature rather than seeking out modern understanding and interpretation. Classic texts such as *Aerodynamics* and *Aerodonetics* (Lanchester 1907, 1908), *Stability in Aviation* (Bryan 1911), *Aeroplane Mathematics* (Brodetsky 1921a), *Dynamical Stability of Aeroplanes* (Hunsaker 1916), and *The Mathematics of Aerodynamics* (Wilson 1918) all help with orientation. In fact, mathematician and physicist George H. Bryan's book played a significant role in the construction of the first stable aircraft, and engineer Frederick Lanchester's weighty tomes also proved to be of importance: contributions that are described later in this book.

One interesting and early technical development seen at the primary research establishments was the wind channel, or what we would now call the wind tunnel. Takehiko Hashimoto's *The Wind Tunnel and the Emergence of Aeronautical Research in Britain* (2000) is very insightful in this respect. Backed up by his thesis of 1990, 'Theory, experiment, and design practice: the formation of aeronautical research, 1909–1930', Hashimoto's modern-era assessments proved to be essential background reading. Structural analysis is also an important area of engineering relating to aircraft strength, and a more reflective book by aeronautical engineer Nicholas J. Hoff, *Early History of Aircraft Structures* (1946), is illuminating.

Regarding where the development of the mathematics of aeronautical engineering could be found in the literature, one would expect to find extensive mathematical discussions in official technical publications, particularly those of the Advisory Committee for Aeronautics (the main policy and control authority in British aviation at the time), and in books dedicated to the subject. Perhaps the most surprising sources of detailed information, however, were the various related journals and general periodicals of the day. A significant portion of this book is dedicated to exploring the role that various manifestations of aeronautical literature played in disseminating information and advancing aeronautics in Britain.

During my research, a number of personal archives came to light that contained all manner of germane literature: private letters, personal diaries, flying logbooks, technical reports, blueprints, and newspaper clippings. It is perhaps here, among these precious collections, where the very essence of this narrative is defined. The material corroborates stories found in other sources, provides deeper insight into certain events, and throws up new, often startling, information. Most importantly, these archives act as a time machine, transporting one back to the period under scrutiny, not only illuminating things based on mathematics but also describing the very nature of life itself in those early days of aviation.

I wanted this book to sit firmly within the history of mathematics and]make its content accessible to as many potential readers as possible. With this in mind, much of the deeper mathematics has been relegated to an appendix, but some basic aerodynamic principles and concepts are explained within the main text where they are required to support the thrust of the narrative. The overarching intent of this book is to illuminate the contributions to early British aviation of many mathematicians, scientists, and engineers whose names have been lost or blurred over time. Some became better known and recognized for contributions in different fields, while others simply never received the accolades they deserved. I also wanted to emphasize the humanity in the story.

The broad structure of the narrative is chronological and takes examples from contemporary literature to illustrate the sort of mathematics in play at the time. It is hoped that by the end of the book readers will have a much better understanding and appreciation of those involved in developing aeronautics in Britain during the early 1900s, some knowledge of the parts played in the story by various forms of literature, and an awareness of the significance of what British mathematicians, engineers, and scientists achieved during the birth of powered aviation.

Abbreviations

AC	aerodynamic centre
ACA	Advisory Committee for Aeronautics
ARC	Aeronautical Research Committee
BA	Bachelor of Arts
BAAS	British Association for the Advancement of Science
BSc	Bachelor of Science
CFS	Central Flying School
CG	centre of gravity
CP	centre of pressure
DSc	Doctor of Science
ELC	East London College
FRS	Fellow of the Royal Society
ICM	International Congress of Mathematicians
MA	Master of Arts
NAL	National Aerospace Library
NPL	National Physical Laboratory
PAPI	precision approach path indicator
PhD	Doctor of Philosophy
QMUL	Queen Mary University London
R&D	research and development
R&M	Reports and Memoranda
RCS	Royal College of Science
RFC	Royal Flying Corps
RMA	Royal Military Academy
TES	*Times Engineering Supplement*
UCL	University College London
UVA	ultraviolet A

THE FLYING MATHEMATICIANS OF WORLD WAR I

1

Introduction

My first solo flight in an aircraft – what an experience! I remember the day as if it were yesterday, but it was actually 6 October 1983. It was Flight Lieutenant Frank Cooper's idea. He was a lovely chap and an excellent flight instructor. Frank had watched me fumble my way around the circuit at Royal Air Force Church Fenton for three-quarters of an hour, and he seemed reasonably convinced I could manage a lap without him sat next to me in the ageing Mk3 Jet Provost trainer. We taxied around to the air traffic control tower and Frank put the safety pins in his ejection seat, undid his straps, pulled the canopy back, and jumped onto the wing. 'Don't crash,' he said, with a big grin on his face. I pulled the canopy closed and watched my mentor walk across the tarmac and disappear into the glass-topped building. There I was, on my own in a jet aircraft. I had a total of 12 hours and 35 minutes of flying experience under my belt and suddenly couldn't even remember how to release the parking brake! The ten minutes that followed will forever be etched on my mind. I remember taxiing around to the end of the south-westerly runway and a calm air traffic officer clearing me to take off. The mighty Armstrong Siddeley Viper turbojet engine roared as I throttled up, and I was soon racing down the runway. The lads on the senior course jokingly referred to the Provost's power plant as the 'variable noise, constant thrust device', but it was certainly producing sufficient power for my purposes at that moment. I lurched airborne, and as the tarmac fell away beneath me the beautiful vista of the North Yorkshire countryside appeared. It felt amazing. I smiled, and as the landing gear clunked into its recesses I levelled off and began the turn downwind, pondering the events that had brought me to this surreal and exciting moment.

A few months earlier I had been innocently looking through the window of the Royal Air Force careers information office in Cardiff High Street watching footage of Royal Navy Sea Harriers in action on a small television. It was the spring of 1982 and the Falklands War was gripping the nation, although I was engrossed in final examination preparations that were proving to be an annoying distraction. A rather heavy hand suddenly fell on my shoulder, and a gruff voice enquired, 'look appealing to you, young man?' I found myself being enticed and cajoled into the building by a burly, uniformed Royal Air Force sergeant, and we were soon embroiled in a deep discussion about my life history and future aspirations. I already had a job lined up – to work as an engineer in the nuclear industry – but life was about to take an unexpected turn. I had been fascinated in the late 1960s and early 1970s by the Apollo space programme, and like many youngsters of my generation I had been seduced by the notion of becoming an astronaut. The British manned space programme had yet to materialize but there suddenly seemed a faint hope that if it ever did, an engineering graduate who knew a bit about atomic energy and who could fly an aircraft might just get somewhere near the front of the queue of potential candidates. A while later, on that crisp autumn day in 1983, instead of being ensconced in the reactor hall of a nuclear power station, I thus found myself alone and airborne in a military jet.

As I drifted back into the moment, the reality of my predicament hit me. Getting off the ground at Church Fenton had been a relatively straightforward affair: wait to reach a suitable speed and then pull the stick back. Landing, however, demanded a more precise fusion of cunning and coordination and, if truth be told, a large dose of good fortune. Frank had made my number one priority explicit – not to crash – but I felt that a reasonable secondary goal was to try to ensure the aircraft survived the impact of my landing without requiring major engineering rectification before its next sortie!

As I turned onto the final approach, all four precision approach path indicator (PAPI) lights positioned next to the threshold of the runway were glowing bright red, and the disconcerting phrase I had been taught to remind me of the implications of this flashed through my mind: 'All four red, you'll soon be dead!'[1] I took in

[1] PAPIs comprise a set of four lights next to a runway threshold that offer a visual indication of the vertical accuracy of an approach relative to a defined optimum flight path. On a perfect approach, a pilot will see two red lights and two white lights. If all the lights appear white, the aircraft is well above the optimum vertical profile; conversely, if all the lights are red, it is well below the profile.

the positives: I was still alive and I had at least remembered to lower the landing gear and extend the wing flaps, so the problem had been reduced to a straightforward battle between gravity, the Viper's maximum thrust, and my (lack of) skill. The calm voice appeared in my ears once again and I was cleared to land. I am sure the PAPIs were turning green at this point, not because they had a third, verdigris filter, but because I was now viewing them through the long grass short of the runway threshold! I exaggerate, but it was certainly a huge relief when tarmac began to rush underneath the nose. I pulled the power off and flared, but slightly too soon – a technique I perfected during my subsequent career, as many of the passengers I had the pleasure of carrying on board various aircraft over the years would testify. The speed washed off rapidly and as the wings decided they had done quite enough work for the day, the aircraft fell with little dignity from the sky onto its main wheels, bouncing slightly before settling onto the hard surface. A voice in my headset congratulated me on surviving the ordeal, and a vision of Frank wiping sweat from his brow appeared. Had he noticed the slightly low approach and bounce, or had his eyes been closed during those final moments, praying to some higher being that his faith in me had not been misplaced and he still had a career as a flight instructor? I turned off the runway at a suitable exit and, after loosening my oxygen mask, heaved a huge sigh of relief. I had made it – I was a jet pilot!

This story recounts a pivotal event in my personal struggle to aviate, but the human quest to master flight has for centuries provided a tantalizing challenge for those inclined to take it on. At the very heart of the endeavour lie those who wrestle with its theoretical and practical aspects: the aeronautical mathematicians and engineers. The mathematics of Archimedes combined with eighteenth-century materials and construction techniques proved sufficient resources to allow development of the first balloons to lift men and women into the air, but controlled, fixed-wing, powered flight presented obstacles that would require a much deeper amalgamation of disciplines and technology to surmount. There was a natural evolution that played out through the whole of the nineteenth century, culminating in the now-famous flights of the Wright brothers on 17 December 1903. A frantic race to develop faster, stronger, safer, more agile and longer-range machines ensued. Driven initially by entrepreneurial industrialists seeking profit and notoriety and later by the more exacting demands of the military, the first two decades of the twentieth century saw exponential growth in almost every aspect of aeronautics. The concomitant burgeoning of the mathematics and engineering underpinning it all is a story in itself: a tale littered with the names of often poorly acknowledged individuals whose efforts fuelled and sustained the startling advancements. This book identifies some of these influential characters and discusses their contributions to aviation in the setting of the research and academic institutions that were most involved in the development of aeronautical mathematics in Britain during the period 1880–1920. Our primary aim is to understand something of their respective backgrounds, expertise, personalities, and interactions, and to chart the contemporaneous technical progress in the field of aeronautics during the four decades astride 1900.

It is reasonable to ask what this book might bring to the table that is unique, given that there are already many excellent works relating to aeronautics during this specific period. It is inevitable that there will be some measure of overlap with other texts in order to provide the necessary background detail to give prominent events and protagonists sufficient context. The intent here, however, is to look beyond the generic landscape and use new, previously unexplored archive material to illuminate and expand upon certain aspects of the existing narrative.

Additionally, in viewing this material through the lens of an experienced aviator, the book attempts to offer a unique perspective. Previous accounts of the period have been written predominantly by historians, whereas this one is crafted through the eyes of a pilot and mathematician. It is hoped that this more intimate connection with the material will provide a conduit to more intuitive and empathetic analysis and commentary. There are many differences when one compares twenty-first-century aviation with that of the early 1900s, yet a plethora of underlying similarities also exist, not least in the attitudes and personal qualities of those closely involved.

Aeronautics is a much-visited area of research and a topic that has fascinated and engaged the minds of many. The landscape of the associated literature is therefore as diverse as the aircraft and people that punctuate its progress, but this wealth and variety of relevant information has the potential to be both a blessing to any prospective author and an impediment. It is helpful to have such a wide selection of source material to draw on when constructing the foundations and framework to support a coherent account, but it is often challenging to identify which material is of most importance. Indeed, it is of genuine concern that something or someone of consequence may be overlooked among the sheer volume of books, journals, articles, and reports pertaining to aviation.

It is also slightly perplexing to identify where best to begin such a story. In searching for the first mention of aeronautics in a British journal, however, happenstance intervened and led me to an article in the first edition of a publication called *Mechanic's Magazine*, which not only discussed the first recorded attempt at human flight in Britain but also had an uncanny personal affiliation. Consideration of this article therefore stood out as a natural way to introduce matters since it conveniently connects the salient themes that permeate the book: aerodynamic challenges, aviation and aviators, mathematics and engineering, aeronautical literature, and acts of incredible bravery in the face of adversity. As an opening gambit, therefore, we meet Eilmer the Flying Monk of Malmesbury, whose alleged aeronautical feat of constructing a pair of wings and launching himself off the top of the local abbey to test them encapsulates and reflects most of these topics and traits.

Understanding the basic challenges facing Eilmer, and indeed all of those who followed him in the pursuit of controlled flight, is crucial if one is to appreciate fully the contributions of those who eventually discovered the pathways that nurtured progress. Around the year 1900, power-to-weight ratio remained a significant hurdle, as did constructing a vehicle that was impervious to the destabilizing effects of wind gusts. The way the aerofoil generated lift remained a mystery – ignorance that had implications relating to structural integrity and control – and technology had some ground to make up to provide suitably resilient construction materials for the proposed new genre of aerial craft.

In determining the cast of characters for this account, two groups stood out: the so-called Chudleigh lot, who were based at the Royal Aircraft Factory at Farnborough; and a group working on stress analysis at the British Admiralty. Both of these collaborations came to light via intriguing photographs published in articles in the centenary edition of the Royal Aeronautical Society's *Aeronautical Journal*: see the images in figures 1.1 and 1.2.[2] Preliminary

[2]See Stephens (1966, 77) and Chitty (1966, 67).

Figure 1.1 The 'Chudleigh lot'.

investigations left many questions unanswered, particularly the identity and roles of the women in the photograph taken at the Admiralty. Were they mathematicians or engineers, or were they present in some other capacity? This was certainly a potentially fruitful avenue of research. It transpired that the Chudleigh group had been considered to some extent by both June Barrow-Green (Barrow-Green 2014) and David Bloor (Bloor 2011, 309). That said, some of the men had received far less attention than others, so here was another possible line of inquiry.

Discussion of the forces that brought these disparate characters together at Farnborough and the Admiralty, and the individual contributions to aeronautics that were made by them, provides the backbone of the book. The story behind the women mathematicians and their associates at the Admiralty also links aeronautical activity to some of the societal changes and challenges that were being experienced and confronted in Britain during World War I. To lay the foundations for all of this analysis it is important to define the state of play in British aeronautics as the country entered the war. The initial discussion thus offers some of the historical background

Figure 1.2 The Admiralty Air Department, 1918.

preceding the conflict. A number of influential individuals are introduced who played significant roles in establishing the mathematics that would eventually underpin the engineering and aerodynamic concepts required for the construction of the first generation of operational flying machines in Britain. Relevant documents and recollections from the personal archives of some of those involved are also integrated into the story wherever possible, these troves containing all manner of pertinent literature and information.

Aside from the Admiralty in London, much of the activity in British aeronautics was focused at two primary research establishments: the National Physical Laboratory (NPL) and the Royal Aircraft Factory. The NPL was established in 1900 at Bushy House, Teddington, to set standards in science and engineering. Various rumblings and statements of intent had been made at a number of British Association for the Advancement of Science (BAAS) meetings during the 1890s as unease grew at Germany's advantage in this field, and the arrival of the British equivalent of the Physikalisch-Technische Reichsanstalt, established in the mid-1880s, was therefore long overdue. The likes of Werner von Siemens and his successors had been driving Germany forward for well over a decade, and it was time for a British response. Initially, the technical institutions and societies dictated the specializations at the NPL: mechanics and engineering, electricity, optics, chemistry,

Figure 1.3 The army airship Gamma II on the ground in front of the airship sheds at Farnborough.

metrology, terrestrial magnetism, and thermometry were the chosen demarcations.[3] As time passed, however, its remit broadened, and in 1909 the aeronautical division was created.

The Royal Aircraft Factory at Farnborough opened for business slightly before the NPL did, but under a different guise. Government-sponsored ballooning began in Britain in 1878, with Captain James Templer given the job of developing and coordinating efforts for the military. Templer was a product of Cambridge University and another character with a notable link to Malmesbury. In 1881 he mounted a balloon sortie from Bath to Dorset, and on board as a passenger was Walter Powell, then the Member of Parliament for Malmesbury (figure 1.4). Events took a turn for the worse as they approached the coast. Templer, attempting a rapid descent to prevent a potential ditching, managed to land the craft just short of the cliffs and jumped clear. Unfortunately, the loss of his weight in the basket caused the balloon to rise again and, despite his best efforts to cling on to the valve line, the buoyancy force won the battle, taking the hapless Powell up and out over the English Channel: Powell was never seen or heard from again.

[3]The nineteenth century saw the founding of a number of professional trades associations, with the Institution of Civil Engineers (1818), the Institution of Mechanical Engineers (1847), and the Institution of Electrical Engineers (1889) no doubt the most influential in determining the NPL's initial technical structure and focus.

Figure 1.4 James Templer and Walter Powell.

Ballooning facilities were initially based in Aldershot, but the successful deployment of balloons as elevated observational platforms in the Boer War, and subsequent rumours of expansion of resources and capability, prompted a move to the less built-up area of Farnborough, the new base becoming operational in 1905. By this time, the Wrights had already designed and flown in a fixed-wing aircraft and airships were very much in vogue, so the days of manned balloons being used as significant instruments of war were numbered.

Perhaps the importance of the Farnborough facility, then, was not so much the balloons it accommodated but more the infrastructure put in place at the location by default just in time to meet the fundamental requirements of a research establishment for fixed-wing operations. What had made the area so attractive in the first place was the space available to accommodate the huge sheds required to house the new British airships that were being planned; it was just good fortune that the surrounding land would prove to be perfect for runway operations. Hence the Royal Balloon Factory was soon transformed and expanded to cater for the new fixed-wing contraptions, morphing into the Royal Aircraft Factory, or simply 'the Factory', as it soon became known colloquially.[4]

In many ways, the NPL and the Royal Aircraft Factory represent the engine in this story. They provided the real estate, the workshop facilities, the funding, and the administration – all the practical necessities required to support the endeavour of meaningful R&D in aeronautics. They were the foci for effort – magnets that would pull together all the necessary resources, both material and human, to make things happen – and represented a concentration of forces that would stimulate and promote rapid progress in this emergent field.

The stage is thus set to tell something of the story of the mathematicians and engineers who would be responsible for the genesis and early development of British aeronautics. It is a vast topic, and this account will only skim its surface, but it is hoped that it will nevertheless offer a sense of what life must have been like for those daring mathematicians in their revolutionary flying machines.

[4]Journalist John Pudney's book, *Laboratory of the Air* (1948), offers a concise description of the founding and evolution of the Royal Aircraft Factory.

2

A Flying Monk

EILMER

In his inaugural address to the London Mechanics' Institution in February 1824, academic philanthropist Dr George Birkbeck described *Mechanic's Magazine* as 'the most valuable gift which [the] Hand of Science has yet offered to the artisan' (Robertson 1824). The first edition of this technical journal appeared on 30 August 1823, and its rather eclectic but easily read palette of articles and correspondence was designed to attract the interest of the contemporary blue collar workers who represented a significant proportion of the market for this genre of literature. The magazine's editor was J.C. Robertson, who was closely linked to the founding of the London Mechanics' Institution.[1] The publication was full of practical advice, interspersed with mathematical puzzles and related commentary. Regarding mathematics-based articles, the focus tended to be on geometry and practical arithmetic, but number theory, trigonometry, calculus, and various other mathematical topics sporadically appeared. Serendipitously, the first edition included an article on aeronautics entitled 'Flying in the air' (figure 2.2). The content of the article turned out to be of interest on many levels, not least because its second paragraph mentions the incredible exploits of Eilmer, the astrologer and flying monk, who, despite the portent, 'could not foresee the breaking of his own legs, for soaring too high' (Anon 1823, 9)!

For many years I have lived next to Malmesbury Abbey (figure 2.3), the very building from which Eilmer reputedly launched himself in the year 1010, his life literally hanging on the efficacy of

[1]The institution and its relationship with Robertson were discussed by Helen Flexner in her doctoral thesis (Flexner 2014).

Figure 2.1 Eilmer the Flying Monk.

his hand-crafted wings. Needless to say, once airborne, his downward velocity rather exceeded his horizontal, causing significant bodily trauma upon his reacquaintance with terra firma. The earliest surviving account of the incident was written more than a century after the flight allegedly took place, so scholars today debate whether it actually happened in the way described. Regardless, the (possibly apocryphal) aviator was known to be more of a mystic than a mathematician, but mention of him in the first issue of *Mechanic's Magazine* made me think this was perhaps a natural and appropriate place to begin a journey that attempts to identify and link interesting characters and feats associated with the early development of British aeronautics and aeronautical engineering.

While Eilmer may not have been a mathematician, the article in question does go on to mention one: John Baptist Dante of Perugia. His flights over Lake Trasimeno using his version of Eilmer-type wings were more extensive and more successful than the British contender's, but, predictably, they also concluded with a rather serious leg fracture. (The injury did not, though, prevent Dante from

MECHANIC'S MAGAZINE.

FLYING IN THE AIR.

Though the science of aerostation is of very modern date, yet there is every reason to believe it was not

at Constantinople, we are told by Knolles, that "amongst the quaint devices of many for solemnizing so great a triumph, there was an active Turk, who had openly given it out,

Figure 2.2 *Mechanic's Magazine*: 'Flying in the Air', 1823.

taking the position of professor of mathematics in Venice on recovery.) In fact, the article does not spare the reader from the physical traumas associated with these brave – some might say 'foolhardy' – exploits as it draws to a close with tales of a certain Mr Murray's attempt at parachuting off the bell tower of Chichester Cathedral in 1790.[2] We hear, rather graphically, of the somewhat disastrous consequences of an inopportune gust of wind as 'the blood gushed from his ears, nose, and mouth very plentifully, and he was many hours insensible' (Anon 1823, 10). The relevance of wind gusts to the stability of fixed-wing aircraft would tease some of the keenest minds in aeronautics a century or so later, so Murray's experience is in many ways prophetic of the first major mathematical challenge that would be presented to those engaged in early aeronautical endeavour.

[2]Chichester Cathedral is the seat of the Anglican Bishop of Chichester and is located in the English county of Sussex.

Figure 2.3 Malmesbury Abbey in 2017.

Mechanic's Magazine was also instrumental in bringing the ongoing work of George Cayley into the public domain. Much of the discussion that follows concerns the work of mathematicians and scientists who somehow morphed into permanent or temporary aeronautical engineers, and there is reasonable justification to nominate Cayley as the first of this breed, certainly in Britain. His pedigree as a mathematician is difficult to establish, but one could reasonably argue that his schooling by the Reverend George Walker and Cayley's demonstrated competence in practical engineering indicate that he at least had a firm grasp of the applied side of mathematics. Walker, elected a Fellow of the Royal Society (FRS) in 1771, was a mathematician and theologian and was Cayley's tutor during the latter's late-teenage years. George Cayley, a distant cousin of famed British mathematician Arthur Cayley, subsequently married Walker's daughter Sarah in 1795.[3]

Cayley's aeronautical exploits can be traced back into the final decade of the eighteenth century. As early as 1799 he defined the

[3]Publications by polymath historian of aeronautics Charles Gibbs-Smith (1965) and engineer J.A.D. Ackroyd (2002a; 2002b) delve more deeply into Cayley's life and highlight his engineering talents.

Figure 2.4 George Cayley.

fundamental configuration of modern aircraft, realizing that the lifting force of the wings had to support the weight of the craft, while the thrust of the power plant had to overcome the drag of the air resistance. His first aeronautically themed publication of note, however, was his three-part text *On Aerial Navigation* (Cayley 1810), which was serialized in *Nicholson's Journal of Natural Philosophy, Chemistry, and the Arts*. Cayley's attempt in 1853 to launch his footman, John Appleby, into the air using a small glider

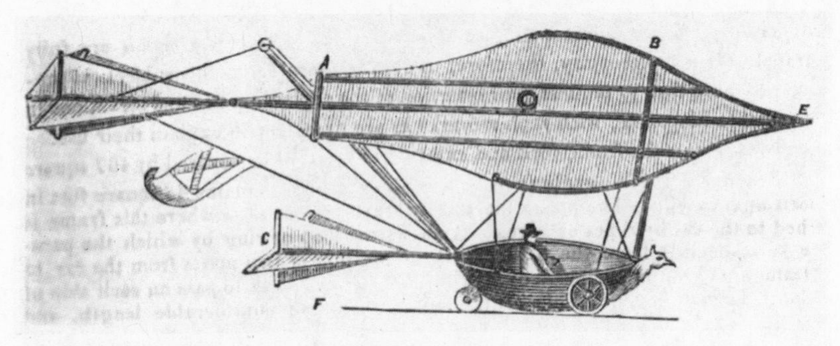

Figure 2.5 Cayley's 'governable parachute'.

being towed down a hill by a posse of volunteers is recognized as a significant milestone in the record of manned flight.

Cayley constructed his first full-size glider in 1849, initially testing it using ballast and then with a servant's ten-year-old son aboard. A modified triplane variant provided the transport that took his footman some 900 feet across Brompton Dale, near Scarborough, Yorkshire, in 1853. There is actually some debate surrounding the identity of the 'pilot' that day, but there is no doubt that the flight took place, with details of the 'governable parachute' appearing on the front page of *Mechanics' Magazine* in September 1852 (figure 2.5).[4]

A couple of years prior to Cayley's exploits of 1849, a paper by the Reverend Alexander Crawford was carried by a rival journal, *The Mathematician* (figure 2.6). It was not directed specifically at aeronautical applications, but considered the motion of a system of particles. Mention of d'Alembert's principle does hint at the sort of analysis that Cambridge Wranglers Edward Routh and George Bryan would undertake later that century and in the early part of the next to define the design criteria to ensure aircraft stability.

THE STABILITY CONUNDRUM

The 'stability' of an aircraft in the sense used here relates to intrinsic aerodynamic properties that tend to dampen or exacerbate the oscillations that result from a disturbance to straight and level flight caused either by certain atmospheric phenomena such as

[4] *Mechanic's Magazine* became *Mechanics' Magazine* from 2 October 1824 onwards.

PROPOSITIONS CONCERNING THE MOTION OF A SYSTEM OF PARTICLES,

SHEWING THAT D'ALEMBERT'S PRINCIPLE IS NOT NECESSARY TO THE THEORY OF DYNAMICS.

[*The Rev. Alexander Q. G. Crawfurd, M.A., of Jesus College, Cambridge.*]

Let m, m', m'', *etc.* be a collection of material particles, any how connected with one another; let P, P′, P″, *etc.* denote the fórces applied to these particles, and a, b, c, a', b', c', a'', b'', c'', *etc.* the angles which the directions of P, P′, P″, *etc.* make with the rectangular axes of x, y, and z respectively: let (m, m'), (m, m''), (m', m''), *etc.* denote the distances of m from m', m from m'', m' from m'', *etc.*: let $[m, m']$, $[m, m'']$, $[m' m'']$, *etc.* denote the forces exerted by m on m', by m on m'', by m' on m'', *etc.*: and lastly, let x, y, z, x', y', z', x'', y'', z'', *etc.* be the coordinates of m, m', m'', *etc.*

By the third law of motion we are entitled to assume, that m' will exert on m a force equal and contrary to $[m, m']$, that m'' will exert on m a force equal and contrary to $[m', m'']$, that m'' will exert on m' a force equal and contrary to $[m', m'']$, *etc.*; and that the forces which each particle exerts on any other is in the direction of the line joining those particles.

By the principles of analytical geometry, we have for the cosines of the angles which the line (m, m') makes with the axes of x, y, and z, respectively, the expressions

$$\frac{x-x'}{(m, m')}, \quad \frac{y-y'}{(m, m')}, \quad \frac{z-z_{\prime}}{(m, m')};$$

and for the other lines which join particles we have similar expressions.

From this notation and these principles it is easy to form the equations of of motion of each of the particles m, m', m'', *etc.*

The equations of motion of m will be

$$m\frac{d^2x}{dt^2} = \mathrm{P}\cos a - [m, m']\frac{x-x'}{(m, m')} - [m, m'']\frac{x-x''}{(m, m'')}, \text{ etc.}$$

$$m\frac{d^2y}{dt^2} = \mathrm{P}\cos b - [m, m']\frac{y-y'}{(m, m')} - [m, m'']\frac{y-y''}{(m, m'')}, \text{ etc.}$$

$$m\frac{d^2z}{dt^2} = \mathrm{P}\cos c - [m, m']\frac{z-z'}{(m, m')} - [m, m'']\frac{z-z''}{(m, m'')}, \text{ etc.}$$

The equations of motion of m' will be

$$m'\frac{d^2x'}{dt^2} = \mathrm{P}'\cos a' + [m, m']\frac{x-x'}{(m, m')} - [m', m'']\frac{x'-x''}{(m', m'')}, \text{ etc.}$$

$$m'\frac{d^2y'}{dt^2} = \mathrm{P}'\cos b' + [m, m']\frac{y-y'}{(m, m')} - [m', m'']\frac{y'-y''}{(m', m'')}, \text{ etc.}$$

$$m'\frac{d^2z'}{dt^2} = \mathrm{P}'\cos c' + [m, m']\frac{z-z'}{(m, m')} - [m', m'')\frac{z'-z''}{(m', m'')}, \text{ etc.}$$

The equations of motion of m'' will be

$$m''\frac{d^2x''}{dt^2} = \mathrm{P}''\cos a'' + [m, m'']\frac{x-x''}{(m, m'')} + [m', m'']\frac{x'-x''}{(m', m'')}, \text{ etc.}$$

$$m''\frac{d^2y''}{dt^2} = \mathrm{P}''\cos b'' + [m, m'']\frac{y-y'''}{(m, m'')} + [m', m'']\frac{y'-y''}{(m', m'')}, \text{ etc.}$$

$$m''\frac{d^2z''}{dt^2} = \mathrm{P}''\cos c'' + [m, m'']\frac{z-z''}{(m, m'')} + [m', m'']\frac{z'-z''}{(m', m'')}, \text{ etc.}$$

If we now add together the first equations of each set, all the terms depending on the mutual action disappear, and the resulting equation is simply

$$\Sigma\, m\frac{d^2x}{dt^2} = \Sigma\, \mathrm{P}\cos a.$$

In like manner by adding the second and third equations of each set, we obtain the results

Figure 2.6 Crawford's *Propositions Concerning the Motion of a System of Particles*, 1847.

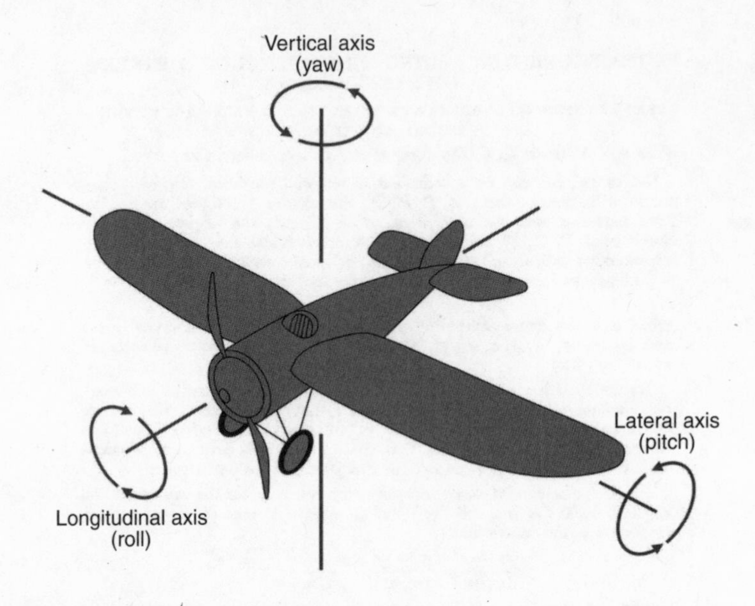

Figure 2.7 Aircraft axes for pitch, roll, and yaw.

gusts of wind or by a pilot's control input via the flight controls. Fundamentally, stability impacts an aircraft's controllability and manoeuvrability in the atmosphere; constructing a vehicle that can be powered into the air is one thing, but being able to safely control its flight path once it is airborne is quite another. To appreciate the primary concerns here, it helps to be familiar with the three main axes about which an aircraft rotates in flight and to consider the intricate dance that is performed between an aircraft's lift and weight vectors in relation to these axes. Figure 2.7 shows the three aircraft axes in the lateral, longitudinal, and vertical planes and the respective rotations about them, known as pitch, roll, and yaw. What follows over the next few pages is not intended as a comprehensive guide to aerodynamic principles and terminology, but more a discussion that offers sufficient information to allow some intuition to be gained regarding the fundamental aerodynamic hurdle facing the mathematicians and engineers in those early days of aircraft design and construction.

Any aerofoil wing produces lift due to the surrounding pressure field it generates as it moves through the air – a field that varies in its distribution with the angle of attack. Angle of attack, or α ('alpha') as it is commonly known in the world of aerodynamics, is

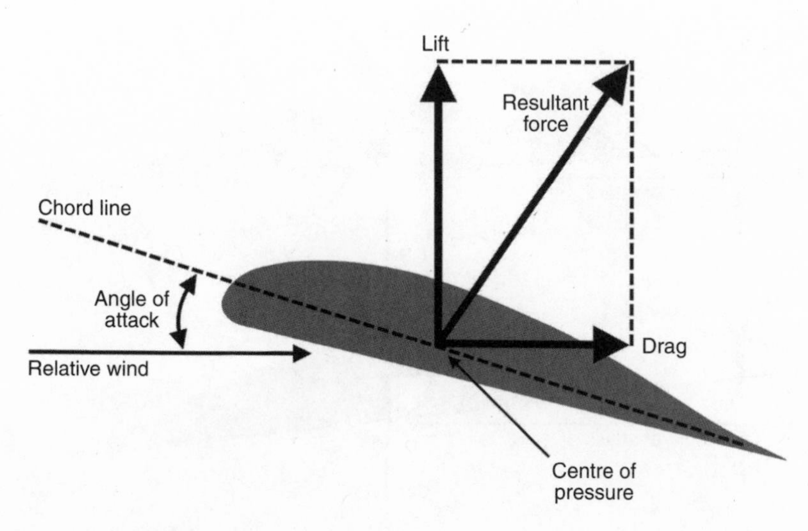

Figure 2.8 Centre of pressure.

the angle made between the wing's chord line and the relative wind direction, as shown in figure 2.8. The relative wind direction is a vector that indicates the relative motion between the wing and the air through which it is moving. The wing will also produce drag by virtue of the mechanism at play that creates the lift. This drag, often referred to as 'lift-induced drag', is related to the energy contained in the vortices shed by the wings. The vortices are formed by higher-pressure air on the lower surface of the wing that is moving in an outward, span-wise direction, spilling around the wing tip towards the lower-pressure region on top of the wing and combining with the chord-wise flow of air over the wing created by the forward motion of the aircraft.

Weight acts vertically downwards through an aircraft's centre of gravity (CG). One can imagine the CG to be the location within the airframe from which one would have to suspend the aircraft for it to hang in perfect balance. Figure 2.8 depicts an aerofoil in flight and highlights a position known as the 'centre of pressure' (CP). We can see that it is the point on the wing through which the lift vector acts. In many ways, the CP is something that might be considered analogous to the CG, but for lift rather than weight. The crucial difference, however, is that any change in α will rapidly alter the pressure field around the wing, which will in turn alter the

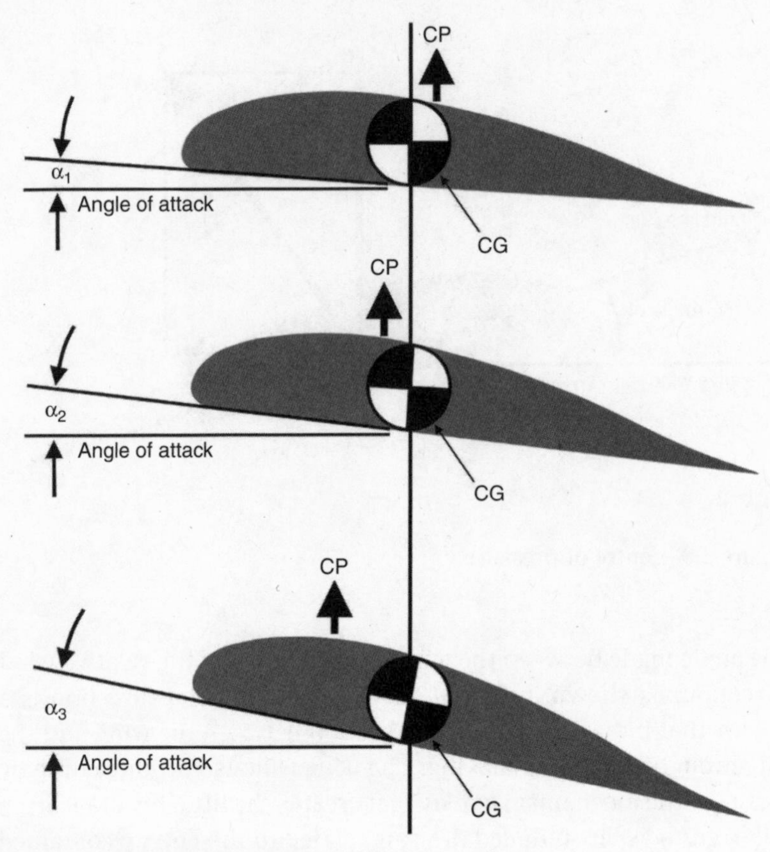

Figure 2.9 Variation of centre of pressure with α.

position of the CP, while the CG ordinarily remains fixed over the short term.[5]

Figure 2.9 shows the same aerofoil at increasing angles of attack, α_1, α_2, and α_3, all below what is known as the 'stall angle'. It is important to understand, here, that a stall in the aerodynamic sense is something very different to the stall of, say, a motor car engine. An aerodynamic stall is not a phenomenon directly related to an aircraft's power plant: it is something that refers entirely to the state of airflow over the wing. In normal flight the airflow adheres closely to the contour of the wing surfaces. When an aerodynamic

[5]Clearly, over an extended period of time, as fuel is burned, the CG will move, but within well-defined limits. Rapid changes in CG tend to be associated only with the dropping of stores, such as munitions or supplies.

stall occurs, the airflow across the top surface of a wing in a sense loses its grip and detaches. When this happens it causes a huge disruption to the pressure field and an almost complete loss of lift. The thing to take away from figure 2.9 is that within the range of α where the airflow remains firmly attached to the top surface of the wing, any variation in the angle of attack will cause the CP to move.[6]

Consider, then, the impact of a moving CP on the longitudinal stability of an aircraft. In normal circumstances, increasing α causes the CP to move forwards, while reducing α results in a rearward movement, as seen in figure 2.9. Any forces with a vertical component that are displaced from the CG in the fore/aft plane will create some sort of rotational force (torque) about the CG, which will manifest itself as a pitching moment on the aircraft.[7] If the CP of the main wing is acting through the CG, there will be zero pitching moment because there is no moment arm; as soon as this CP moves away from the CG, however, some pitch will be induced. To counter this and keep the aircraft in level flight, a balancing equal and opposite moment has to be generated, and this is provided by the horizontal stabilizer at the tail, which is invariably also an aerofoil. This adjustment can be made using moveable surfaces attached to the rear of the stabilizer (elevators) or by having some mechanism in place to be able to change the α of the horizontal stabilizer itself in flight. By modifying the lift generated at the tail, the effect of a moving CP on the main wing can thus be controlled. Longitudinal stability fundamentally depends, therefore, on the interplay between the main-wing CP and the aircraft CG, and on the compensatory moments generated by the tailplane assembly. Figure 2.10, a silhouette of a Bristol M1c aircraft in level flight, illustrates a typical situation. In this example, the main wing lift is acting ahead of the aircraft's CG with a moment arm **x** to produce a pitch-up rotation. Since the aircraft is flying straight and level, its horizontal stabilizer must be providing enough lift acting on moment arm **y** to create an equal but opposite pitch-down rotation.

This aerodynamic interaction along the longitudinal axis of the aircraft is exacerbated by the intrinsic aerodynamic twisting forces experienced by the aerofoil itself. Aerofoils are designed in such a

[6]This effect can clearly be seen in figure 3.11 on p. 59: an image of a stalled aerofoil during a wind tunnel test at the NPL during World War I.
[7]For a detailed definition of 'torque', see p. 236.

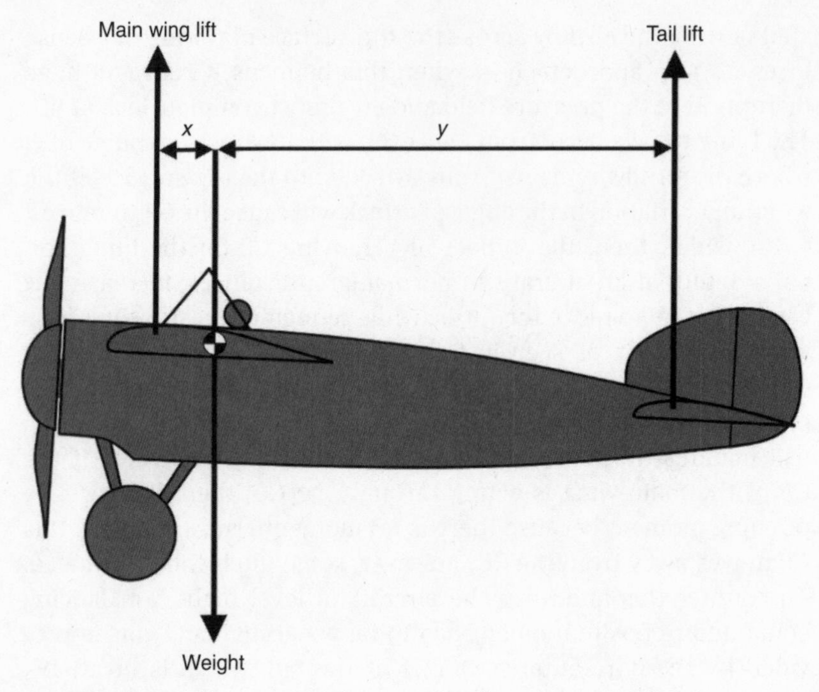

Figure 2.10 Stabilizing, opposing rotational moments in pitch.

way that there is a particular point, known as the 'aerodynamic centre' (AC), about which pitching moments produced by the changing pressure field remain constant. In other words, as α changes, the position of the CP changes, and so too does the magnitude of the lift vector, but in such a way as to ensure that the net moment about the AC remains constant. Despite being constant and predictable, however, this is yet another rotational force acting in pitch that needs to be calculated and considered in any aircraft design process. Figure 2.11 shows an aerofoil wing in flight, increasing its angle of attack from α_1 to α_2. The original anticlockwise moment about the AC was generated as $(L_{\mathrm{CP1}})(d_1)$. At the new angle of attack, the pressure field change causes the CP to move forward. However, the net moment about the AC does not change, since $(L_{\mathrm{CP2}})(d_2)$ has the same magnitude, and acts in the same sense, as the previous moment $(L_{\mathrm{CP1}})(d_1)$.

Perhaps we can now better appreciate one of the most pressing dilemmas facing early aircraft engineers: what configuration of

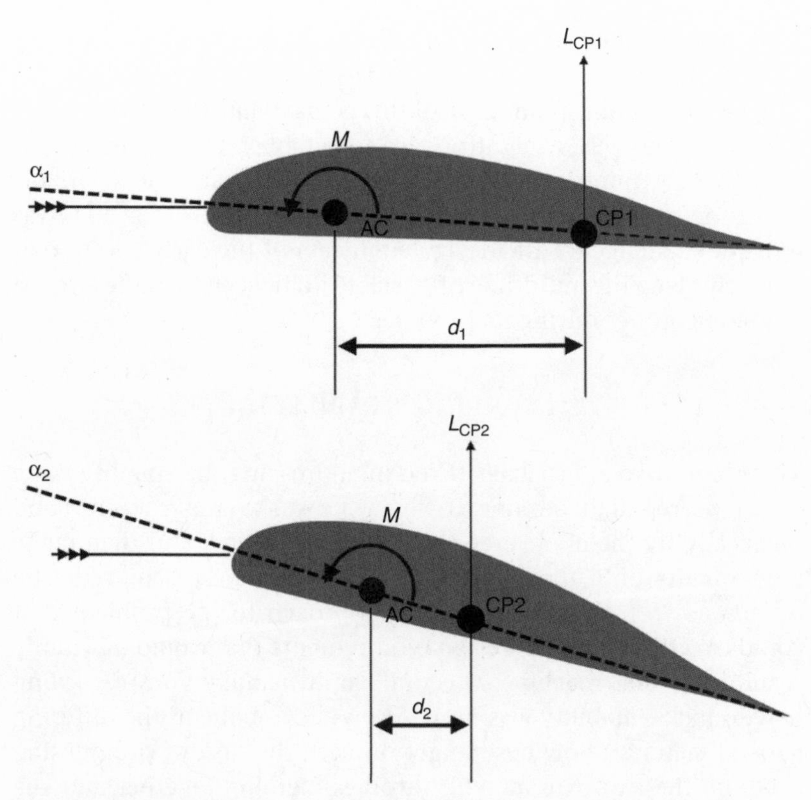

Figure 2.11 Constant rotational moment, M, about the aerodynamic centre with changing α.

main wing and tailplane results in an aircraft that is inherently stable in pitch? In other words, what design features would ensure any movement of the main wing CP in relation to the aircraft CG caused by arbitrary disturbances would not result in a loss of control? Similar challenges existed in the other two axes of the aircraft, of course: designs had to ensure inherent stability in roll and yaw. A relatively simple solution to the former involved adjusting the dihedral of the main wings to give the desired effect. The dihedral is the upward fixed angle of an aircraft wing from the horizontal. It tends to make an aircraft more stable in roll by increasing the lift on the down-going wing while reducing it on the up-going wing during any roll. Yaw was a less troublesome issue since any rotation about the vertical axis of the aircraft tends to be self-correcting. The rudder presents a larger face to the oncoming airflow as yaw is applied or induced: an effect that naturally causes a correcting yawing moment.

The overarching point in all of this is that while the CG of any aircraft was relatively straightforward to establish for a given configuration, determining the relationship between it and the invariably poorly defined lift vectors being produced by the aerofoil wings was a major challenge for the design engineers of the early 1900s, particularly given the multitude of possible flight scenarios the aircraft might encounter during its lifetime.

TACKLING STABILITY

There were two approaches taken in addressing the stability issue once powered flight became a reality. One was to construct craft and physically fly them, observe how they performed, and then make adjustments until the desired characteristics were achieved. The other was to pursue a theoretical approach to the problem that could exactly define the design requirements that would inevitably result in a stable machine. This philosophical dichotomy regarding how to tackle stability was in many ways symbolic of the differing general views on how best to get on with the task of progressing aviation: the two extremes of 'throw something together and see what happens' versus the more measured strategy of 'do the mathematics first, build the craft later'. In exploring the origins and progression of the latter approach, the importance of the contributions of a number of nineteenth-century academics is apparent.

Edward Routh became one of the most iconic and well-respected figures among the many outstanding Cambridge University mathematicians during the second half of the nineteenth century. This most prolific of coaches would prepare hundreds of candidates for the arduous Mathematical Tripos examinations and become a legend in his own lifetime during the process. It is rather ironic that his death in 1907 coincided with some of the very first attempts in Britain to master the controlled flight of fixed-wing, powered flying machines. He can justifiably be considered as one of those sitting at the very top of the tree of British characters who influenced the metamorphosis of certain elements of the mathematics that formed the backbone of aeronautical engineering. He not only provided a summary of the core mathematics required to characterize and define stability in a general sense, but also influenced the

Figure 2.12 Edward Routh.

mathematical thinking of the man who would subsequently develop the theory in the specific context of aviation: George Hartley Bryan. Bryan was not the only important link in the chain following Routh either. Another of Routh's pupils, John Strutt (later Lord Rayleigh), would also play a significant role in the aeronautical policymaking that would dictate the early course of aeronautical research in Britain when he became chairman of the Advisory Committee for Aeronautics (ACA) in 1909.[8]

[8]Of the mathematicians mentioned in this narrative, in addition to Bryan and Strutt, Routh coached Bertram Hopkinson, Horace Lamb, Micaiah J.M. Hill, Joseph J. Thomson, and Arthur Berry. He also coached Robert A. Herman, who himself would go on to coach Selig Brodetsky. Chapter 5 of Warwick (2003, 227–85) provides detail of Routh's teaching. The ACA is discussed in more detail in chapter 3 of this book.

Figure 2.13 George Stokes.

The story of the quest for an understanding of stability in dynamic systems is deeply rooted in the nineteenth century. Routh's initial contribution to the required mathematics, *A Treatise on the Stability of a Given State of Motion* (Routh 1877b), was a response to the Adams Prize question posed collectively by James Clerk Maxwell, James Challis, and George Stokes for the period 1875–1877, which demanded an essay that addressed what it called 'The criterion of dynamical stability'. Stokes himself made significant advances in the field of fluid dynamics, particularly during the 1840s, with the Navier–Stokes equations providing essential insight into the nature of air flows.[9] Across la Manche in 1854, the French mathematician Charles Hermite had established some solid foundations in a letter to German mathematician Carl Borchardt, in which he alluded to the analysis of the roots of an equation of quadratic form to determine stability characteristics.[10] Routh was unaware of Hermite's remarks, however, instead drawing upon assertions made by Augustin-Louis Cauchy (Routh 1877b, 24–5).

[9]They are still used today in aerodynamic contexts and analysis, though the proof regarding the existence and uniqueness of solutions to the equations remains one of the unsolved Clay Mathematics Institute's Millennium Problems. In the year 2000, the Clay Mathematics Institute stated seven 'Millennium Prize Problems' in mathematics, with a correct solution to any of them attracting a prize of US$1 million.

[10]Hermite is perhaps best known for his proof that e (the base of natural logarithms) is a transcendental number. He and Borchardt met in Paris in 1846 while the former was working at the École Polytechnique and the latter in Königsburg.

As mentioned, the stability conundrum specifically relating to aircraft would eventually be comprehensively addressed and solved by mathematician and physicist George Bryan. Bryan graduated from Peterhouse, Cambridge, as 5th Wrangler in 1886 following his thorough grounding in mathematics from Routh. In 1895 he moved across into a teaching and research post as professor of pure and applied mathematics at the University College of North Wales in Bangor. His passion was more inclined towards thermodynamics and statistical physics, but these were subjects that brought him together with Austrian physicist and philosopher Ludwig Boltzmann, who had among his friends both German gliding pioneer Otto Lilienthal and American aviation stalwart Samuel Pierpont Langley.[11] It would be this circle of aeronautical enthusiasts that would draw Bryan into aviation and eventually convince him to tackle the stability issue. Chapter 5 of Routh's *An Elementary Treatise on the Dynamics of a System of Rigid Bodies* (Routh 1877a) provided the necessary mathematical framework for the work.

While similar attempts were being made across the Channel to harness stability, Bryan was able to derive a more advanced theory than the likes of Ferdinand Ferber, Marcel Brillouin, Rodolphe Soreau, and Hans Reissner. Brillouin had developed a static theory rather than a dynamic one, while Soreau failed to account for the interdependence of yaw and roll in aircraft motion in his analysis.[12] Bryan's work was unashamedly and firmly constructed around the dynamic theory of rigid bodies developed by Routh and, to some extent (and probably begrudgingly, given Bryan's contempt for practical engineers), the work of versatile engineer Frederick Lanchester.[13]

[11]Physicist T.J.M. Boyd offers an excellent account of Bryan's journey from Cambridge into the world of aviation. He was a latter-day successor of Bryan at the University of North Wales, Bangor, and his *One Hundred Years of G.H. Bryan's Stability in Aviation* (Boyd 2011) tells the whole story of how this seminal work came into being.

[12]The static theory of Brillouin, a mathematician and physicist educated at École Normale Supérieure in Paris, only considered conditions for the stable equilibrium of the aircraft without any external disturbances such as gusts of wind. During the early 1900s, Brillouin was professor of mathematical physics at the Collège de France. Soreau was influential in nomographic as well as aviation circles. Reissner was central to the German efforts to drive progress in aeronautics, his most notable achievement being the design of a controllable-pitch propeller, although there is some doubt if it was ever used in flight (Kinney 2017, 105).

[13]Frederick Lanchester is discussed in chapter 3.

Figure 2.14 George Bryan.

Bryan's work culminated in 1911 with the publication of his book, *Stability in Aviation* (1911), which is arguably one of the most important theoretical texts written in pre-war Britain in the field of aeronautics. The book's contents were actually an extension of a paper Bryan had written with William E. Williams in 1904 (Bryan and Williams 1904), the theory therein relying on knowledge of certain parameters that Bryan called 'resistance derivatives'; nine had to be determined for longitudinal stability, and a further nine for lateral stability. A more detailed description of these parameters and a closer look at Bryan's mathematics can be found in the appendix. It is sufficient to say here that these numbers defined specific stability equations and were primarily functions of air resistance encountered in flight, and as such they needed to be physically measured in the air rather than calculated in some theoretical sense on the ground. In the book's preamble its author's frustration at the paucity of academic resources being employed to solve the theoretical problems surrounding the propensity, or otherwise, of aircraft to depart stable flight is very clear (Bryan 1911, 16–17):

> As a branch of higher applied mathematics the study of aeroplane motions has sadly been neglected. The vaguest notions still prevail even as to the very meaning of stability ... It is only by the co-operation of University teachers in mathematics and physics that the present anomalous conditions can be remedied.

Despite Bryan's annoyance, his book of 1911 would become the keystone in the quest to produce the first inherently stable aircraft in Britain, although the culmination of this endeavour came a couple of years later, after its publication.

THE AERONAUTICAL SOCIETY OF GREAT BRITAIN AND ITS JOURNAL

While the foundations for the mathematics of aeronautical engineering were being laid during the second half of the nineteenth and early twentieth centuries, interest in aeronautics was growing in Britain, exemplified by the founding of the Aeronautical Society of Great Britain and the proliferation of articles in respected journals such as *Nature*. The Aeronautical Society was founded in 1866 and had the objectives of advancing 'aerial navigation' and observing any developments in what it called 'aerology'.[14] At its first meeting, Francis Wenham delivered a lecture entitled 'Aerial locomotion and the laws by which heavy bodies impelled through the air are sustained'. Wenham was a marine engineer with a keen interest in aviation, and he spent much of his time working on

[14]The founding members were the Duke of Argyll, James Glaisher, Hugh Diamond, Francis Wenham, James Butler, and Fredrick Brearey.

the design of gliders that incorporated superposed wings: multiple, short, stacked wings that could generate the equivalent lift of a single, long wing. This, indeed, was a design principle that was embraced during the early part of the 1900s in the biplanes and triplanes that were ubiquitous in the skies during the period. One of Wenham's most important contributions to the development of aeronautics in Britain was his part, along with colleague John Browning, in the design and construction of the first wind tunnel, completed in 1871. This advance added a new dimension to the concept of testing aircraft designs and gathering data, and it was the precursor to the more powerful and robust tunnels that would become an essential component of the R&D at the Royal Aircraft Factory and the NPL a few decades later. In many ways, Wenham took up the reins from George Cayley and was the man who helped guide and nurture British aeronautics into the new century.

The Aeronautical Society played its part in this, and its role was enhanced by the introduction, towards the end of the 1800s, of its official mouthpiece: *The Aeronautical Journal*, which became central in disseminating news to the close-knit aeronautical community of the time regarding progress in the field. It was a quarterly publication that first appeared in January 1897, and in the third issue, in July of that year, mathematics applicable to aeronautics made its first appearance, including Duchemin's formula for the magnitude of the air force on a flat plate;[15] indeed, by the fourth issue, readers were being exposed to whole pages of mathematics. The man behind the journal's instigation was Captain Baden Baden-Powell, the society's honorary secretary.[16] His father, Baden Powell, was an accomplished mathematician who held the Savilian Chair of Geometry at Oxford from 1827 to 1860. The journal would gain significant momentum through the early years of the twentieth century under the editorship of John Henry Ledeboer – graduate of Caius College, Cambridge, who had studied medieval and modern

[15]Duchemin was a colonel in the French army, and he carried out the experiments to determine his formula in 1829. The formula predicts the aerodynamic force, P, on an inclined surface, with inclination α degrees to the horizontal flow, when the force on the surface when perpendicular to the flow, P', is known: $P = P'(2\sin(\alpha)/(1+\sin(\alpha)))$.

[16]Baden Baden-Powell was the younger brother of Robert Baden-Powell, founder of the Boy Scout Movement.

Council, 1914-15.

The Council for the year 1914-15 is composed as follows:—Associate Fellows: A. E. Berriman, Griffith Brewer, Harris Booth, J. H. Ledeboer, Lieutenant A. R. Low, R.N.V.R., Squadron Commander F. K. McClean, R.N.A.S., Squadron Commander Alec Ogilvie, R.N.A.S., Mervyn O'Gorman, C.B., F. Handley Page, Dr. T. E. Stanton, Lieutenant-Colonel F. H. Sykes, Dr. A. P. Thurston. Members: Squadron Commander G. Aldwell, R.N., Colonel H. E. Rawson, C.B., R.E., Major-General R. M. Ruck, C.B., R.E., Dr. R. Mullineux Walmsley.

Figure 2.15 Council of the Aeronautical Society, 1914–15.

languages at university but whose real passion was aeronautics – assisted by Baden Baden-Powell himself, who sat as president of the society from 1900 to 1907. It is interesting to note how the council of the society acted as a magnet to attract many of the individuals discussed in this story into a common forum; figure 2.15 shows a list of the council members for 1914–15, for example. We see, here, evidence of how the pieces of the aeronautical jigsaw fitted together at this time. There are representatives from the press (Berriman and Ledeboer), the research establishments (O'Gorman and Stanton), academia (Thurston), the military (Ogilvie), and industry (Handley Page). There is no doubt that the members of the council held great sway in determining the direction of British aeronautics in the eras before and after World War I, although the primary rudder remained the ACA and its derivative, the Aeronautical Research Committee (ARC).[17]

NATURE

The subject of flight is mentioned in *Nature* in an 1870 issue, the reference being to balloons in France. It would be the exploits of Hiram Maxim, American-born British inventor and entrepreneur, however, that in 1895 would attract the first report linking aeronautics and mathematicians. Maxim is probably best known as the inventor of the Maxim gun, the first portable fully automatic machine gun,[18] but it was his spectacular demonstration of a potential flying vehicle using an aircraft 'of sorts' powered by steam and running on a rail that caught the headlines on this occasion.

[17] The ACA became the ARC in 1920: a change in name rather than function.

[18] Maxim's gun was developed to become the Vickers machine gun, a weapon that was standard issue for the British armed forces for many years.

Figure 2.16 Maxim's aircraft on rails, 1895.

George Greenhill, 2nd Wrangler in 1870 and professor of mathematics at the Royal Military Academy (RMA) in Woolwich, was the author, and not only did readers get the benefit of his erudite musings, they were also treated to photographs capturing the historic event, as seen in figures 2.16 and 2.18. The article was devoid of mathematics, but once fixed-wing, powered flight had begun in earnest, *Nature* did not shy away from introducing the public to the mathematics of aeronautics.

Maxim's exploits are little known these days, even among contemporary aviators, so it is worth expanding on this event and others associated with him. A first-hand account is available in the Presidential Address given to the Newcomen Society in 1949 by Dr Albert Thurston, entitled 'Reminiscences of early aviation' (Thurston 1949).[19] Thurston, who was Maxim's engineer, tells us that the flying machine weighed 10,000 pounds and had a wingspan of 104 feet; it was powered by two steam engines, each developing 180 horsepower.

[19]The Newcomen Society was formed in Birmingham, England, by a number of enthusiasts, with the intention of bringing together people interested in the history of engineering.

Figure 2.17 Hiram Maxim.

There is a tongue-in-cheek claim that this machine achieved the first powered 'free flight', the large construction only being prevented from lifting off by the guide rail to which it was attached. The date was 31 July 1894, the location Bexleyheath, in southeast London.[20] Thurston's association with Maxim was strengthened when the former was appointed chief assistant and designer, and the pair set to work on a commercial venture to bring 'the experience of flight' to the general public using what amounted to fairground rides. Thurston's recollection of the first test ride is as worrying as it is hilarious (Thurston 1949, 2):

> After it had been tested for strength by means of sand bags, the engineer-in-charge thought to give the young designer [Thurston] a thoroughly enjoyable run. Speed was increased until the centrifugal force reached 6.47 times gravity. After a mighty mental struggle I fell unconscious to the bottom of the cab. Fortunately I had with me in the machine a tough workman who was in a more advantageous position, and he was able to give the signal to stop.

[20]Thurston's personal records indicate that the incident may have occurred the following day, on 1 August 1894.

Figure 2.18 Maxim's entourage, 1895.

I can relate only too well to Thurston's predicament. I came to regard aerobatics as a fine 'spectator sport'. It all seems very spectacular and such fun to the innocent onlooker, but when actually sat in the cockpit during the twists, turns, and loops, the associated g-forces give a pilot a very different perspective and physical experience. Perhaps the best way to put it is that it was an aspect of my flying training that I coped with rather than enjoyed. I remember Ted Vary, my kindly squadron boss at Church Fenton, describing with a wry smile on his face my attempt at an aerobatic display sequence – which he experienced from the other seat in the aircraft – as 'agricultural but effective'. I think we both knew I was never going to be on the shortlist of prospective candidates to join the Royal Air Force Red Arrows elite aerobatic display team!

Returning to Thurston's musings, a comment he makes that hints at the primal wrangling between practical engineers and the theoreticians appears later on in his piece. He states that, 'In the early days of aeronautics, pure mathematics was as great a barrier to progress as the sonic barrier has been recently' (Thurston 1949, 5). It is clear from this article that Maxim taught the young Thurston much, including how to find the centre of pressure of an aircraft wing, how to protect fabrics against actinic (UVA) rays, and how to promote automatic stability, to name but a few things.

CHANUTE

As the nineteenth century drew to a close, aeronautical engineering could hardly be described as a mature or expansive discipline. Perhaps the book that best summarized the progress to that point was Octave Chanute's offering of 1894: *Progress in Flying Machines* (Chanute 1894). Chanute, born in Paris in 1832, had moved to the United States at the age of six, and during his formative years he developed a keen interest in civil engineering, a field in which he would excel throughout his working life. His interest in aviation came during his later years, however, when he set out to consolidate all that was known about experimental aeronautics, a pursuit that resulted in his book. If there was a 'standard pre-1900 text', this was probably it. In fewer than a dozen of the early pages Chanute covers the main points known in aerodynamic theory at that time. In fact, it was from Chanute's book that the term 'aeronautical engineer' entered the vernacular. The book discusses many other aspects of the story of aeronautics and includes some mathematical arguments and expressions, Duchemin's formula being typical. A useful appraisal and summary of Chanute's work is given by John D. Anderson Jr in his comprehensive reference book *A History of Aerodynamics* (Anderson 1997, 192–7).

A new century dawned, and it was very much a case of the calm before the storm. Strands of mathematics, practical engineering, industry, entrepreneurship, literature, experimentation, and desire were all being woven together. From these threads the fabric of aeronautical engineering would emerge in earnest to provide the reassurances needed to construct, power, and safely operate flying machines. Such flight was still not a reality, but conceptually it was no longer in the realm of pure fantasy.

3

Overture to War

TRIALS AND TRIBULATIONS

Britain had invested heavily in balloon technology around the turn of the century, meaning that the basic infrastructure was in place to facilitate a rapid move towards the new genre of aerial craft. One man, John Dunne, stood out as the person given most responsibility to oversee and coordinate the major aspects of this transition, and it is through his efforts and interactions that we can monitor and chart aeronautical progress in Britain through the first decade of the 1900s. Without doubt, Dunne's first priority was to configure an aircraft that could at least perform in a comparable way to other craft that were appearing in various countries around the world. At this point, the largest stumbling blocks remained achieving a sufficient power-to-weight ratio and guaranteeing stability and control.

Many people are familiar with the Wright brothers' flights of late 1903 in the United States: exploits that are generally recognized as the first controlled, powered flights in a fixed-wing aircraft. In Britain, however, initial attempts to emulate the feat did not take place until somewhat later, and few people would be able to name the individuals and aircraft involved, or the location of their (as it transpired, futile) initial endeavour. The first bona fide British trials were carried out covertly in Scotland, coordinated by Dunne, a second lieutenant in the British Army who had been sidelined with illness but had put his time in convalescence to good use contemplating potential aircraft structures for the military.[1] Dunne's aircraft designs invariably incorporated the distinctive V-shaped

[1] A comprehensive account of Dunne's role in the trials, and a view on the man himself, can be found in chapters 7–9 of aircraft design engineer Percy B. Walker's second volume of *Early Aviation at Farnborough* (Walker 1974), in which he discusses the first aeroplanes to appear at the Royal Aircraft Factory.

Figure 3.1 J.W. Dunne.

wings that were in many ways prophetic of modern-day lifting surfaces. His D.5 aircraft outline, in particular, resonates with that of many early jet-powered aircraft and, of course, with the Northrop Grumman B2, the twenty-first century's stealth bomber. Although well ahead of its time, the full significance of the swept-wing concept would not have been obvious to Dunne until much later in his life. Modern aircraft designs often employ swept wings to delay the onset of shock waves associated with exceeding the sound barrier, for example – not something that would have been a consideration for aircraft or pilots in the early 1900s!

His talent for aeronautical innovation did not go unnoticed, and he soon found himself posted to Farnborough as the military hierarchy woke up to the fact that there might be some tactical advantage to be gained from command of the skies. And this is how Dunne and his select entourage eventually came to find themselves on a bleak hillside in Scotland in 1907 attempting to emulate the Wright brothers.

Dunne would prove to be somewhat of a linchpin during the period that might be considered the equivalent of the 'Big Bang' in British aeronautics. Not only would he play a key role in these first British trials of fixed-wing, powered machines, he would also become an influential and highly regarded figure among those who were intimately involved in the new discipline. He sat on the most important committees and became the go-to man for the aeronautical press whenever an informed quote or opinion was required. David Edgerton would have it that Dunne 'shamelessly exploited his family connections with the Army to keep his experiments going'

(Edgerton 1991, 3), and while there may be a large element of truth in this view, Dunne's genuine talents as an aeronautical engineer are hard to dispute.[2]

His industry and status meant that he generated and received a significant amount of correspondence. Fortunately, many of these original exchanges survive in the form of letters, notes, and telegrams, revealing a who's who of key players in contemporary British aeronautics, politics, the military, and the media.[3] His first appearance in Farnborough was in 1905, after spending the previous two years experimenting with model gliders. He was given the task of coordinating tests on full-scale aircraft with a view to producing a design that demonstrated inherent stability. By 1907 – which, incidentally, was the same year construction of the first wind tunnel was completed at the Royal Aircraft Factory – Dunne had a pair of 12-horsepower Buchet engines at his disposal and a construction he believed would actually fly: the D.1-A airframe. A marriage of the two resulted in the D.1-B, the first iteration of a series of powered aeroplanes that Dunne would mastermind.

A decision was made in Whitehall to keep the D.1 trials as secret as possible, so a small detachment of hand-picked personnel discretely left Hampshire bound for the remote Glen Tilt, Blair Atholl, Scotland. Although initially clandestine, these attempts at controlled flight eventually received significant coverage in newspapers and periodicals, particularly in *Flight*, in which photographs and a descriptive narrative appeared in 1910. A retrospective and more comprehensive account appeared in the same publication many years later, in 1943 (figure 3.2). In fact, a short article did escape censorship and feature in the *Automotor Journal* in 1907, but it offered scant detail.

Unfortunately, things did not go quite to plan with the trials in 1907, much to the chagrin of Richard Haldane, British Minister for War, who had made the long journey north from London to bear witness to what he thought would be a landmark achievement. Round two of the trials, in 1908, proved more fruitful, however, with Lieutenant Lancelot Gibbs making a brief flight in the D.4, which was a significantly improved version of the D-1 series of aircraft designs.

[2]Dunne was clearly comfortable with design and practical engineering, but discovering the major influences that fostered such competencies has proved challenging. There is little evidence to suggest he was a mathematician.

[3]The Dunne archive is a recent acquisition of London's Science Museum.

556 FLIGHT MAY 27TH, 194

A DUNNE BIPLANE OF 1913: The panels between the wing tips were fins only and were not used as rudders, all controlling being done by the trailing-edge flaps. The castor oil smoke from the 50 h.p. Gnome engine brings back early Hendon memories. The undercarriage was by way of being a "tricycle."

Tailless Trials

Tribute to a British Pioneer: The Dunne Biplanes and Monoplan

By C. M. POULSEN

TO the modern generation Col. J. W. Dunne is known as the author of books such as "An Experiment with Time" and "Nothing Dies," which out-Einsteined Einstein. Only the old-timers realise that he was one of our earliest aircraft pioneers. These notes may help to show how far he was ahead of his time, aeronautically speaking.

GREAT interest has been aroused by the article published in our issue of May 13th, 1943, in which, under the title "Turbines and the Flying Wing," Mr. G. Geoffrey Smith, our Managing Editor, suggested the logical combination of the tailless type of aircraft and power plants of the turbine-compressor type. Briefly, the suggestion was that instead of trying to adapt orthodox monoplanes to the new type of power plant, the aircraft might be designed around the power plant.

Many are the attempts that have been made to produce really practical tailless flying machines. Reference has been made to several of them in our pages in the past few months. Names such as Junkers, Dunne, Hill, Lippisch, Handley Page and Northrop, recall interesting designs. In view of the revival of interest in the tailless and its large-scale version, the "all-wing" type, it has appeared to us not only fitting that a very sincere tribute should be paid to the "father" of all tailless aircraft, Col. J. W. Dunne, but desirable that the results of his early work should again be made available to the aircraft community. It fell to *Flight* to record, from 1910 to 1913 or so, the trials and triumphs of Col. (then Lieut.) J. W. Dunne, but that was before many present readers took any interest in flying!

Wing span - - 36ft.
Wing chord (root) 6ft. 3in.
Wing chord (tip) 5ft. 0in.
Wing area - 230 sq. ft.
Area of flaps (2) 18 sq. ft.

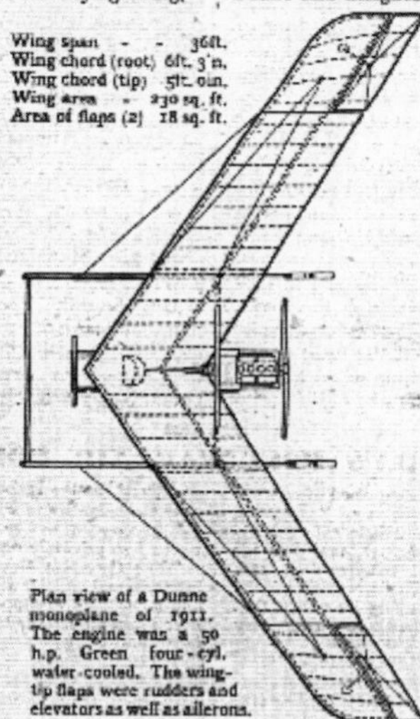

Plan view of a Dunne monoplane of 1911. The engine was a 50 h.p. Green four-cyl. water-cooled. The wing-tip flaps were rudders and elevators as well as ailerons.

It is nothing short of amazing, especially in view of t relatively small progress made since, that as early as 19 Dunne had designed his tailless monoplane wing. In following year he showed it to C Capper, of the Army Balloon S tion (from which later grew t R.F.C., R.N.A.S., and R.A.F Col. Capper, however, asked for biplane, and a biplane version w designed and built at Blair Atho

To appreciate to the full the for sight and insight of J. W. Dun it is necessary to recall that at th time aerodynamic knowledge w almost non-existent, and what lit there was had not been assimilat by the majority of aircraft expe menters. Lanchester understoo from theoretical considerations, t mechanism of lift and drag, but f of those who did the practical e periments were able to follow h reasoning, in detail at any ra "Box Kites" were the order of t day, and speeds were so low th the problem of reducing d scarcely entered into the pictu In this connection one may, p haps, recall as an example of t way in which experimenters w more concerned with structu strength than with aerodynamic r finement, the famous "thru test." It was a common sayi that if, in a biplane structure, th was room for a thrush to fly b tween the bracing wires, anoth

Figure 3.2 Poulsen's tribute to Dunne in 1943.

The Dunne archive contains letters detailing exchanges between the man himself and the editors of the early aeronautical journals: dialogues that offer deep insight into how material moved from the sphere of physical engineering into the realm of the media. There seemed to be a relationship between the technical editors and the aeronautical industrialists that surpassed the need to simply provide news; it was often a two-way exchange of information and opinion, symbiotic in many ways. In a personal letter to Dunne on 17 January 1913, for example, Algernon Berriman, the technical editor of *Flight*, gives his own opinion as to why there had been such reluctance in the mainstream to accept the possible benefits of Dunne's negative wing tip design feature:

> The conditions of the industry have tended, rather, I think, towards the general acceptance of the orthodox in design, just as they did in the early days of motoring, when Lanchester produced an essentially original chassis which the public of the time would not buy because the continental type was regarded as the orthodox thing to have.

The term 'negative wing tip' was used by Dunne to describe a design feature that had the very end of each wing angled downwards to modify control and stability. More modern aircraft today tend to have wing tips angled upwards to improve the lift-to-drag ratio by modifying the wing tip vortices: a concept that was not understood in the early days of flight.

John Ledeboer, who had moved on to become editor of the popular journal *Aeronautics*, was also a frequent correspondent with Dunne on technical matters, but the exchanges were not exclusively of a technical nature. Aviation was somewhat of a political football at the time – something that was obvious from Dunne's dialogue. Charles Grey, the editor of *The Aeroplane* publication, was one of the most vociferous critics of the government's tardy response to the need to define and implement a coherent aerial defence policy for Britain, and he would do anything to canvass support and gain verbal ammunition from the likes of Dunne. In a stratum of more operational significance, Dunne was also a trusted confidant of General Sir David Henderson, Director-General of Military Aeronautics, one of numerous high-ranking military personnel at the War Office with whom the junior officer had dealings.

Dunne can thus be viewed as the hub of a many-spoked wheel, the midfield general of the contemporary aeronautical soccer team.

Sadly, history seems to have relegated this influential pioneer to the substitutes' bench in favour of Samuel Cody, a more glamorous and flamboyant player, who received all the mainstream plaudits and accolades at the time for his pioneering flight on 16 October 1908 – a flight that saw him travel 463 yards across Farnborough's airfield during the maiden voyage of the imaginatively named British Army Aircraft Number One.

Studying figure 3.3 in close detail, the precarious and exposed nature of flight at this time is blindingly apparent. Cody is perched ahead of the power plant (a 'pusher' design) and has no protection from the elements other than his clothing and footwear. The photograph seems to show moveable control surfaces located midway between the upper and lower wings, and one wonders if these were perhaps integrated into the design to circumvent the Wrights' comprehensive patent on wing warping as a means of control. Note, also, that the lifting surface architecture is based more on a Wenham-type arrangement of multiple, short, stacked wings than on a wide single-wing design. This particular flight terminated in a crash, as did many at the time, and it is truly remarkable that any of these pilots survived long enough to learn from their experiences. Cody did eventually succumb to the perils of flight when he was killed on 7 August 1913, his own design for a floatplane breaking up in mid-air during a test flight.

As an addendum, Dunne moved from aeronautics into the realm of philosophy during the interwar years, developing a popular theory describing the nature of time. 'Serialism', as it was known, was fully explained in his book *An Experiment with Time* (Dunne 1927), a publication that, remarkably, brought its author far greater recognition than did any of his work with aeroplanes.

INDUSTRIAL ACADEMIC INFLUENCE

While a small number of thoroughbred eggheads worked on aeronautical issues behind the scenes, some academically minded industrialists took on central roles in developing practical aeronautics in Britain, Frederick Lanchester perhaps being the most prominent. He certainly followed a less spectacular academic pathway than many in this story, but that did not prevent him gaining a sound working knowledge of mathematics and engineering. In

Figure 3.3 Samuel Cody flies the British Army Aircraft Number One at Farnborough, 1908.

1885 he won a national scholarship to the combined Normal School of Science and the Royal College of Mines in London. In 1887 he also attended classes at Finsbury Technical College to accelerate his learning, but he never gained any formal qualifications. This lack of official recognition did not hamper his success as a businessman in the automotive trade, but it would have serious ramifications regarding his credibility among academics, particularly during encounters with those from Cambridge, when debates raged about topics such as how a wing actually generates lift. He was a grass roots engineer who produced a brace of extremely important books on aeronautics. His first work was *Aerodynamics* (Lanchester 1907), which concentrated on his ideas and thoughts regarding lift and drag. His second piece, *Aerodonetics* (Lanchester 1908), dealt primarily with stability. In the latter book, Lanchester coined the term 'phugoid' to describe certain motions he observed during experiments on models.

It is difficult to assess the most important aspect of Lanchester's contribution to British aeronautics. He was an omnipresent character with a multitude of roles and interests. He would doubtless have been influenced by John Perry and William Ayrton while at Finsbury, but for a man without a classical training in advanced mathematics,

Figure 3.4 F.W. Lanchester.

the level of mathematical exposition in his books is extraordinary.[4] Always erring towards the practical, he seemed to have a natural talent for translating what he observed during experiments into applied mathematics on paper.

He did not confine his findings and understanding to his books, either: he was one of those who was happy to write articles for the aeronautical press to enlighten the readership. One such contribution, 'The flying machine from an engineering standpoint' (Lanchester 1914), carried in *Flight* and published in 1914, perhaps epitomizes how the applied mathematics of aeronautics was being expounded to the masses at the time. Based on the notes Lanchester had used in delivering the prestigious James Forrest Lecture to the Institution of Civil Engineers on 5 May 1914, his article, which was distributed over a number of issues of the popular magazine, brings the reader up to date on the state of play of engineering in British aeronautics immediately prior to the start of World War I.[5] In hindsight, it was probably quite a handy summary for those across the Channel with an interest in assessing British capabilities in the

[4]Perry was appointed to be the chair of mechanical engineering at Finsbury Technical College in 1882 before becoming professor of mechanics and mathematics at the Royal College of Science (RCS) in 1896. He was the driving force behind the 'Perry Movement' that championed reform in the mathematical education of engineers during the early 1900s. During his time at Finsbury he collaborated with Ayrton, the Chair of Applied Physics at the college, with whom Perry had worked previously at the University of Tokyo.

[5]James Forrest was the Secretary of the Institution of Civil Engineers for 40 years (1856–1896), and an annual lecture was set up in his honour following his retirement.

air, and it even directed readers to the various associated 'Technical Reports' of the ACA pertinent to the topics under discussion; these reports contained all the detailed mathematical arguments pertaining to the aeronautical research being conducted in Britain at the time.

An aviator might raise a smile at Lanchester's assessment of the relevance of the various degrees of stability being implied by the curves he derived that illustrated the possible flight trajectories of a generic aircraft left to its own devices given various starting parameters: 'From the engineer's point of view it is unimportant whether the flight path stability is inherent in the machine, or whether, so to speak, the finishing touches have to be given by the pilot himself' (Lanchester 1914, 523), he insists. Of course, the concentration demanded of a pilot who has to keep an inherently unstable aircraft under control is enormous compared with that required to safely fly and operate a stable craft! At the very start of the article, Lanchester gives a mathematical equation to describe the flight path of his hypothetical flying machine. The main curves of interest that emerged from his analysis were the phugoids, with the flight paths that resulted in some form of loop 'being only of interest to mathematicians and the student of "trick flying" ', as Lanchester put it.

To be asked to deliver the Forrest Lecture and then summarize its content in *Flight* shows that Lanchester clearly held sway in the aeronautical community, but he nevertheless had to endure criticism from many of his Cambridge-educated colleagues, particularly Bryan. Lanchester was, unwittingly, the closest of all those involved in the field to completing the puzzle of aerofoil lift, yet he was possibly the one most doubted by his peers.[6]

During the war years he somehow found the time to ponder the limitations on the size of aircraft imposed by the potential weight of the required wings. In 1916, he made public his thoughts in an article for *Engineering* (Lanchester 1916), giving a theoretical argument asserting that the weight of wings would increase as a function of $W^{3/2}$ for total weight of aircraft W, assuming one wished to keep the load safety factor the same. This clearly had profound implications for the perceived practicality of constructing much larger, heavier aircraft.

[6]For a full discussion of Lanchester's theory of lift and the controversy it created, see chapter 4 of Bloor (2011).

His work ethic was second to none, so it was a great loss to aeronautics when he moved more towards his automotive interests. It is fitting that he is now remembered as one of Britain's great industrialists, but his impact on the development of early British aircraft is not as widely known or appreciated as it deserves to be.[7]

IMPENDING HOSTILITIES

In 1909, and despite his disappointing experience in Scotland with Dunne, Haldane continued to view developments in aviation with interest, conscious that the revolutionary heavier-than-air machines would undoubtedly be key participants if, as seemed likely, hostilities commenced in Europe. He considered it prudent, therefore, to exercise a measure of control over the rather disparate band of aeronautical protagonists who had emerged during the previous decade. Consequently, on 30 April of that year he formally established the ACA, a body of men that would orchestrate the continuing evolution of British aviation. Louis Blériot's historic flight in his Type XI aircraft across the Channel less than three months later brought Haldane's prescience into sharp focus, as it became apparent to most that Britain could no longer hide behind its impressive navy; to remain oblivious to the threat from the air would indeed be folly.

The potential importance of the ACA was clear, so Haldane knew its leader had to be someone with established credentials; an obvious candidate, a man with gravitas and credibility in abundance, was Lord Rayleigh. He was a heavyweight in contemporary science and mathematics, already regarded as a world authority in the field of acoustics following publication of his two-volume work *The Theory of Sound* (Rayleigh 1877). He was also a Nobel Laureate, receiving the Nobel Prize for Physics in 1904 in recognition of his 1894 discovery of the noble gas argon (see figure 3.5). He had indicated his personal interest in aeronautics as early as 1891, authoring a

[7] In Coventry, a place with strong links to his automotive business, the local College of Technology was named after him, as was its derivative through merger, Lanchester Polytechnic. Sadly, partly due to confusion with other educational establishments with similar-sounding names (such as Manchester), his name was eventually replaced by that of its host city. Coventry University, as the establishment is now known, does at least have a 'Lanchester library'.

review in *Nature* (Rayleigh 1891) of Langley's *Experiments in Aerodynamics* (Langley 1891). His subsequent delivery in Manchester of the Wilde Lecture in 1900, entitled 'The mechanical principles of flight' (Rayleigh 1900), enhanced his aeronautical profile, and this, combined with his experience in hydrodynamics, gave him the ideal résumé.[8]

In fact, Rayleigh would become the driving force behind the discontinuity theory of aerofoil lift, championed by Britain until the post-war era, when the circulatory theory, most associated with the German engineer Ludwig Prandtl, was seen to reflect a much closer representation of reality. The discontinuity theory's origins in this context dated back to Rayleigh's 1876 paper 'On the resistance of fluids', in which, when talking of a flat lamina at an angle to a water flow, he notes that (Rayleigh 1876, 434):

> Behind the lamina there must be a region of dead water bounded by a surface of discontinuity, within which the pressure is the same as if there were no obstacle. On the front face of the lamina there must be an augmentation of pressure.

The circulatory theory of lift argued that the presence of an aerofoil in an airflow induced a circulation of air around it which modified the normal flow in a way that reduced pressure on the upper surface while increasing it on the lower. Lanchester had been arguing the merits of a circulatory component in theoretical lift analysis for years, but this solitary British voice was lost amid the noise in support of Rayleigh's theory. The perceived lack of rigour in Lanchester's mathematics, coupled with his engineering background, made Rayleigh's offering far more plausible to the academics who dominated influence and comprised the consensus.[9]

Someone taking a macroscopic view of aeronautics in Britain at the start of the second decade of the twentieth century would have witnessed a veritable potpourri of influences; that said, broad delineation is possible. Aside from the Balloon Factory and the NPL, industry, academia, and the Admiralty were all salient players. Faced with this rather eclectic fellowship, it made sense for Rayleigh

[8]The Wilde Lecture was named after Henry Wilde, FRS, electrical engineer and president of the Manchester Literary and Philosophical Society, 1894–1896.

[9]See David Bloor's *Enigma of the Aerofoil* (Bloor 2011) for a comprehensive analysis of the contemporary debate surrounding the mechanism of aerofoil lift generation.

Figure 3.5 *Vanity Fair*'s caricature of Lord Rayleigh discovering the inert gas argon.

to source the members of his committee widely. The new Advisory Committee would sit at the heart of British aviation to control, guide, and coordinate efforts until after the war. A closer inspection of the prominent features of the aeronautical landscape at this time reveals the nature of the response to this formalization of intent on behalf of the British government. Aviation, which had traditionally revolved around flying demonstrations, aerial meets, static exhibitions, and pleasure flights, was no longer in the entertainment business: it was being surreptitiously readied for war.

THE RESPONSE OF RESEARCH AND ACADEMIC INSTITUTIONS

THE ROYAL AIRCRAFT FACTORY

The Royal Aircraft Factory, née Balloon Factory, at Farnborough was undoubtedly the most important facility at the disposal of the ACA. As we have seen, the Factory went through something of an identity crisis during the immediate pre-war years. Its main purpose in its guise as the Royal Aircraft Factory, however, was to be a source of innovation in design in aeronautics and a primary research establishment working in parallel with the NPL. Despite the name changes, the balloon facility remained in situ under the supervision of Colonel John Capper, but it was the key appointment by the ACA towards the end of 1909 of Mervyn O'Gorman as superintendent to lead the departure into fixed-wing aircraft development that would prove decisive. An electrical engineer who learnt his trade at the City and Guilds Central Institution in London, O'Gorman brought industrial experience, scientific insight, and clear vision to the role.[10]

The new superintendent faced significant challenges in building the foundations of Britain's primary, fixed-wing, aeronautical

[10]The City and Guilds of London Institute for the Advancement of Technical Education was founded in 1878 following a meeting two years earlier of 16 of the City of London's livery companies. The objectives were to establish a 'Central Institution' in London that would be used to improve the training of craftsmen, engineering technicians, and professional engineers, and to establish a system of qualifying examinations in technical subjects. The Central Institution was eventually opened in 1884 in South Kensington. In the interim, and as a stop gap, Finsbury Technical College was established to perform a similar role in the heart of the furniture-making district of London.

Figure 3.6 Mervyn O'Gorman.

research centre. Fortunately, however, soon after his appointment at Farnborough, help was at hand in the person of design engineer Frederick Green.

Green had cut his teeth working as an engineer for the Daimler Company in the early British motor trade, and he was appointed as Engineer in Charge of Design on the recommendation of Lanchester. His specialization was engine design, and O'Gorman immediately put him in charge of a number of projects: the design of the BETA, GAMMA, and DELTA airships; the S.E.1 and B.E.1 aircraft; and all engine development from the Royal Aircraft Factory-1.a model onwards.[11] A serendipitous reunion at the 1910 Aero Show in London between Green and an old friend, the pioneering aircraft designer and pilot Geoffrey de Havilland, soon brought the latter into the fold, and the new team set to work on improving the performance and design of the first prototype aircraft that Cody had flown at Farnborough in 1908. It was not long before the collaboration had designed, built, and tested a new prototype, and, by March 1912, its second iteration, the B.E.1 (Blériot Experimental 1), emerged; by August that same year, the B.E.2 variant had appeared.

In the summer of 1912, O'Gorman's tour de force was enticing an academic from Cambridge, who had been awarded a degree in the first class in the Mechanical Sciences Tripos and who was also a qualified pilot, to join the group. Edward (Ted) Teshmaker Busk was given the title assistant engineer physicist, and he immediately began coordinating tests on de Havilland's B.E.2a, as it was now designated.

Such was the significance of this appointment and the impact of the cascade of events it triggered that much of chapter 5 of this book will be devoted to Busk and his contributions. It is sufficient to note here that upon his arrival, Busk mainly worked in the office, making sense of the experimental results fed to him by the pilots nominated to fly the test profiles. However, he soon became frustrated by the lack of empathy that these men demonstrated regarding what he was trying to achieve, so in March 1913 he convinced O'Gorman that the only solution was for him to conduct his own experiments.

[11] Green would later head up the physics department at Farnborough. He continued his career in the aeronautics/motor industry with the Siddeley-Deasy Company following his departure from the Royal Aircraft Factory in 1916.

Figure 3.7 Aircraft at Jerseybrow, Farnborough, 1915. This was an engineering facility set up to repair Royal Flying Corps aircraft that had been too badly damaged in action to allow local repair on the front line.

Figure 3.8 Edward Teshmaker Busk in the R.E.1.

De Havilland agreed to take Busk on as his protégé and taught him the rudiments of test flying. As previously described, Bryan had published his seminal work *Stability in Aviation* (Bryan 1911) a year or two before, and Busk was able to take Bryan's theory, understand it, and then set to work producing a design for a stable aircraft. The fruit of his labour was the R.E.1 (Reconnaissance Experimental 1), which he first flew in the spring of 1913 and which is widely considered to be the first inherently stable aircraft ever constructed (figure 3.8); it would be the forerunner of the B.E.2c, one of Britain's aviation workhorses during the early part of World War I (figure 3.9).

It is worth reflecting here that while Busk went about his work, there was still no agreed theory to describe what made these contraptions, held together by bits of string and canvas, fly at all. Somehow he had managed to construct a machine that would not only fly but that also possessed arguably the most important intrinsic characteristic (certainly of a reconnaissance aircraft): that of stability in flight. His impact at the Royal Aircraft Factory went deeper than just masterminding a brilliant feat of engineering, though: he had set a precedent. Here was a Cambridge graduate who had combined his academic prowess and industrial experience with flying skills to produce something of practical worth in the field of aeronautics: a mathematically based academic training melding with genuine engineering.

Prior to leaving the Royal Aircraft Factory at the end of his seven-year contract and spurred on by Busk's precedent, O'Gorman had put together a team capable of unlocking some of the mysteries of flight and advancing knowledge in this fledgling field. Some of

Figure 3.9 The B.E.2c.

this cohort had been drawn into aeronautics on its own merits and would go on to make careers out of it, whereas others were pressed into service by the morbid necessities of war and would return to pursue their true passions once peace had been brokered – those who survived, of course. By the time Henry Fowler, who had been chief mechanical engineer for the Midland Railway, replaced O'Gorman as superintendent in 1916, there were assembled at Farnborough some of Britain's finest mathematical and engineering minds: the 'Chudleigh lot'. Named after the house in which they messed, from this group would emerge the individuals who inspired this book. They were exceptionally brave and academically talented characters who became the epitome and definition of magnificent mathematicians in their flying machines.

Thus, from its humble beginnings, the Royal Aircraft Factory blossomed into the mainstay of the ACA's asset portfolio. Not only did it thrive in aircraft R&D and associated construction, it also engaged in the mass production of its most successful designs, much to the chagrin of the privateers who thought this to be very much encroaching on their territory. Indeed, O'Gorman left the Factory rather abruptly, as he had been held responsible for performance issues with the B.E.2c and also because of the controversy surrounding the establishment's role in the mass production of aircraft and aero engines: after all, the Factory's remit had been to pioneer design through R&D and then delegate any bulk production of aircraft components or complete airframes to the private sector.

The importance of the Factory was also reflected in the size of its workforce, which, by the end of the war, exceeded 5,000, 1,500 of whom were women (see Barrow-Green 2014, 80). The status of, and opportunities for, women in the engineering workplace improved enormously during the war years – a topic explored by many historians of gender and technology – but this positive momentum was stifled in the immediate post-war era.

AERONAUTICS AT THE NATIONAL PHYSICAL LABORATORY

The man given the responsibility for running the NPL prior to and during the war was Richard Glazebrook. The Trinity-educated mathematician and physicist had graduated as 5th Wrangler in 1876 and had close links with Rayleigh through their collaboration at the Cavendish Laboratory, so his appointment to the Bushy House post was no surprise.[12] The new aeronautics division fell under the umbrella of Thomas Stanton's engineering department and was headed by Leonard Bairstow, who was well on his way to establishing himself as one of the most influential and important British mathematicians in early aerodynamic research.[13]

Bairstow had served his advanced mathematics and engineering apprenticeship under John Perry at the RCS in London. He was the leading experimentalist in the field of aircraft stability and a great advocate of the wind tunnel as a source of reliable data. Turning down Glazebrook's offer of the superintendent's position of the proposed new aerodynamics department in 1917, he instead took up a post at the Air Board under Alec Ogilvie, where he engaged

[12]The Cavendish Laboratory opened in 1874 in Cambridge under James Clerk Maxwell. Rayleigh took the helm as Cavendish Professor of Experimental Physics in 1880 following Maxwell's premature death. Glazebrook, appointed as a demonstrator at the laboratory in the same year, hoped to succeed Rayleigh, but J.J. Thomson prevailed in the 1884 election despite Rayleigh's overt support for Glazebrook.

[13]Stanton had been professor of engineering at Bristol University prior to his appointment at the NPL. Harold Roxbee Cox, a post-war expert on aeroelasticity, apparently suggested, in a mischievous moment, that the roof of the large wind tunnel at Farnborough should be decorated with statues of all the great pioneers of aeronautics from 'Leonardo da Vinci to Leonardo da Bairstow' (Temple et al. 1965, 31)!

Figure 3.10 Leonard Bairstow.

in the coordination of work pertaining to the structural strength of aircraft.[14]

Both main research establishments were keen to offer articles to aviation periodicals outlining recent developments. In an extract from one of them, Leonard Bairstow makes observations and comments that spell out the state of aerodynamics by 1913:

> Aeronautics, or rather aerodynamics, is almost entirely an experimental science at present, and a glance backwards through time will show how much the fundamental data of the subject is derived from experiments on models. [So] we must still continue to make experiments with models, and in doing so must ask ourselves whether the flow around models is like that round the flying machines, and, if not, we must know how to find our conversion factors.

He goes on to say that (Bairstow 1913, 330):

> A true theory of aerodynamics would answer those questions for us completely, but unfortunately for us the answers to such questions are beyond the reach of our present mathematical knowledge.

This honest appraisal, slightly biased against the strides made by those flying the real aircraft and gathering data at Farnborough, admits that the properties of the aerofoil remained a mystery, and that how to relate data from the wind tunnels to the flying machines was a tantalizing mathematical enigma.[15] A particular feature of Bairstow's article, which had been presented as a paper to the Aeronautical Society a few weeks before appearing in *Flight*, was the inclusion of actual pictures from the wind tunnel experiments (figures 3.11 and 3.12). These must have been a real novelty and an eye opener for the readership, and they no doubt allowed many to gain some basic insight into the subject without having to negotiate the mathematical hurdles that are erected later in the piece. The realistic worth of his experiments is given with slightly more humility by Bairstow in his final paragraph, where he asserts that 'it appears to be impossible to do anything better than to make predictions from model tests, using such discretion as may be suggested by experience on actual flying machines' (Bairstow 1913, 330). These

[14] Alec Ogilvie's role at the Air Board is discussed on p. 149.

[15] Bairstow also had issues with Lanchester's views of aerofoil lift: see Bloor (2011, 148–53).

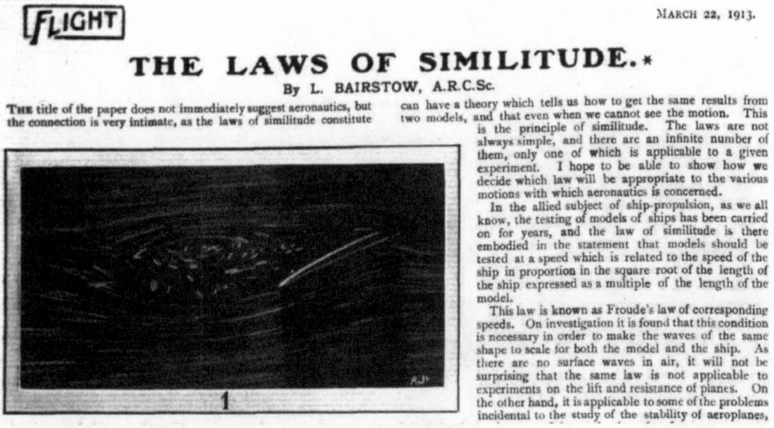

FLIGHT MARCH 22, 1913.

THE LAWS OF SIMILITUDE.*

By L. BAIRSTOW, A.R.C.Sc.

THE title of the paper does not immediately suggest aeronautics, but the connection is very intimate, as the laws of similitude constitute can have a theory which tells us how to get the same results from two models, and that even when we cannot see the motion. This is the principle of similitude. The laws are not always simple, and there are an infinite number of them, only one of which is applicable to a given experiment. I hope to be able to show how we decide which law will be appropriate to the various motions with which aeronautics is concerned.

In the allied subject of ship-propulsion, as we all know, the testing of models of ships has been carried on for years, and the law of similitude is there embodied in the statement that models should be tested at a speed which is related to the speed of the ship in proportion in the square root of the length of the ship expressed as a multiple of the length of the model.

This law is known as Froude's law of corresponding speeds. On investigation it is found that this condition is necessary in order to make the waves of the same shape to scale for both the model and the ship. As there are no surface waves in air, it will not be surprising that the same law is not applicable to experiments on the lift and resistance of planes. On the other hand, it is applicable to some of the problems incidental to the study of the stability of aeroplanes,

1

Figure 3.11 Bairstow's 'The laws of similitude' of 1913.

words would have resonated with any aviator: the academics were doing their best, but any venture skyward was still to some degree a leap of faith. There was potentially as much guesswork as there was hard science underpinning the whole enterprise.

The image contained in figure 3.11 shows a reality which, at that time, eluded detailed explanation and analysis via mathematics. It depicts an aerofoil in a flow (from right to left) of brightly illuminated oil droplets, the photograph having been taken with a camera exposure of one second. The length of each trail thus gives an average velocity of the flow at that point in space. This aerofoil is in a stalled condition, with the turbulent air from the detached flow over the top clearly visible. What Bairstow shows in figure 3.12 is the effect of increasing the flow speed of a fluid passing a flat plate. In images 2 and 3 the fluid is water; in 4 and 5 it is air – again, the flow is right to left. Instead of oil, Bairstow uses Nestlé's milk on the back of the plate in order to see the fluid motion.[16] With low fluid speeds (images 2 and 4) we see a continuous corkscrew sheath in the plate's wake, but as the flow speed increases, at some point the nature of the wake completely changes to defined loops (images 3 and 5). These sorts of experiments and photographic records would eventually give engineers the insight required to begin to understand the nature of wing vortices and fully appreciate what is going

[16]Nestlé's milk was (and still is) a popular, sweetened, tinned condensed milk.

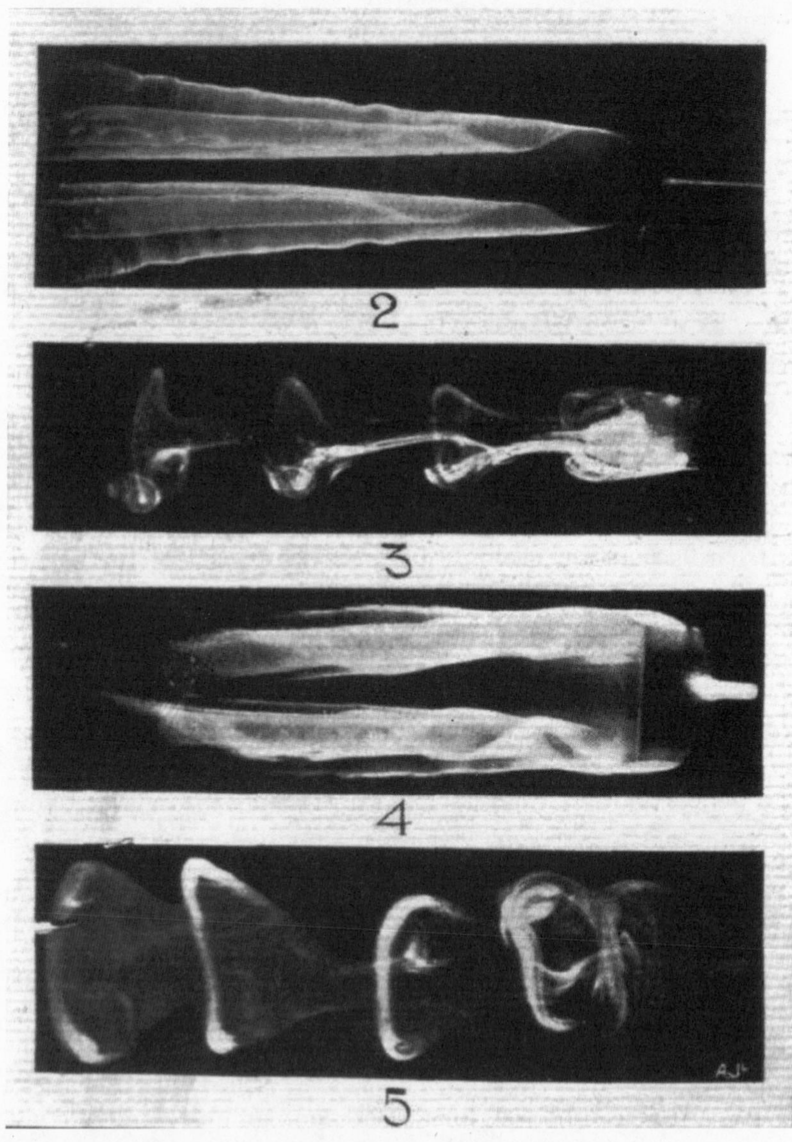

Figure 3.12 Bairstow's images from the NPL wind tunnel.

on with the airflow over an aerofoil during various phases of flight. A picture is perhaps worth a thousand equations when it comes to aerodynamics.

THE UNIVERSITIES

Many of Britain's universities and colleges also became involved in aeronautics as the subject's status within academic circles was promoted during the immediate pre-war period; indeed, as the first decade of the century came to a close, the aeronautical genre was beginning to be recognized in its own right as an important branch of engineering. Imperial College began a scholarship programme allowing talented students of mathematics, engineering, and physics to take on work under Bairstow at the NPL.

Arthur Fage, who arrived at the NPL in 1912, was a typical example of such a student. Fage was one of the so-called navy scholars: characters who attended 'dockyard schools' prior to arriving in research posts. These unique academic institutions were responsible for nurturing and producing some of Britain's finest engineers and scientists. Ernest Relf was another good example. He, like Fage, served his apprenticeship in the Royal Naval Dockyard in Portsmouth between 1904 and 1909 before being awarded a scholarship to the RCS, from where he moved on to the NPL in the pre-war era to gain research experience. As things transpired, he would spend many years at the Teddington facility, becoming superintendent of the aerodynamics division during the 1920s: a post he held until the end of World War II. Relf eventually became principal of Cranfield College of Aeronautics. Fage, who won prizes for his work in mathematics and theoretical fluid mechanics prior to arriving at the RCS, was aided on his journey by John Perry, who, by this time, had moved on from his position at Finsbury.

News of the NPL scholarships, and details of how to apply for them, was given in publications such as *Nature*, which also updated readers on academic courses that were becoming available at institutions such as the Regent Street Polytechnic (Anon 1910). Lectures in aeronautics also began at Imperial College, spearheaded by the high-profile visiting lecturer George Greenhill, who had recently retired from his post at the RMA. His six-lecture series, delivered in March of 1910 and March of 1911, formed the chapters in his book of 1912: *Dynamics of Mechanical Flight* (Greenhill 1912).

Figure 3.13 George Greenhill.

At University College London (UCL) in 1916, as the demands of war permeated through the fabric of British life, Karl Pearson's cell of human computers was eased away from the analysis of national statistics to pursue more war-applicable endeavours. A doyen of mathematical statistics, Pearson was 3rd Wrangler in 1879 and spent much of his academic career at UCL, where he held the chair of applied mathematics and mechanics.[17] Pearson was in direct contact with mathematicians at the Royal Aircraft Factory, who were wrestling with the theoretical aspects of propeller efficiency. Whenever those at the Factory came across an integral that could not be easily solved or evaluated, they immediately turned to Pearson for assistance and advice.[18] Manchester, too, forged ahead with relevant research, particularly in the field of fluid dynamics, with the Osborne Reynolds Laboratory being used exclusively for the inspection and calibration of aircraft instruments throughout the period of conflict. Even Cambridge turned a hesitant gaze skywards.

It would not be until after the war, however, that the universities really started investing heavily in this sector, with Cambridge in particular developing an important school of aviation research under Bennett Melvill Jones. Jones had been a student at Cambridge

[17] See Barrow-Green (2015) for details of the roles of Pearson and his laboratory staff during the war.

[18] Much of Pearson's correspondence with the Royal Aircraft Factory is preserved among his personal papers, which are held in the UCL archives.

of the influential Bertram Hopkinson, and he was also a close friend of Ted Busk. Throughout the war he worked on the development of aircraft instrumentation and gunnery, and his experience and background at the NPL, the Royal Aircraft Factory, and as a pilot led to his appointment as the first Francis Mond Professor of Aeronautical Engineering at Cambridge in 1919.[19]

EAST LONDON COLLEGE AND ALBERT THURSTON

As the NPL set up its dedicated aeronautics group, so too did the East London College (ELC), now Queen Mary University of London (QMUL), under Albert Thurston. Thurston, a graduate in mechanical and electrical engineering from ELC in 1906, returned to his alma mater to establish and orchestrate the first aeronautical engineering department in Britain in 1909. He gave lectures on the subject and is credited with a significant role in the development of the first leading-edge devices for wings: slats, in modern-day parlance. He was awarded his DSc, a qualification introduced in 1860 by London University to reflect advanced study, in recognition of his aeronautical research. Then, after adding a flying licence to his qualifications, he was given responsibility for the safety design requirements and structural integrity testing of military aircraft for the duration of the war that soon followed.

Destructive testing was a technique being used to improve the structural integrity of aircraft, and Thurston played a key role in this aspect of British aircraft design.[20] Original papers exist that detail his team's tests on early aircraft such as the D.H.4 and the Bristol Biplane. It would be reasonable to add Thurston's name to the list of academic aviators who contributed a great deal to British aeronautics yet today have been almost forgotten. Looking through the original notes and sketches he made during the war, it is easy to see that here was a man who had the safety of his fellow aviators at the forefront of his mind. Many more pilots would likely

[19]The Francis Mond Professsorship was named in honour of a Peterhouse graduate who lost his life while flying with the Royal Air Force on the Western Front.

[20]Destructive testing is an exploratory procedure that seeks to determine the strength of a component or structure by imposing ever-increasing loads until physical failure occurs, rendering that component or structure useless following the test. Non-destructive testing, on the other hand, seeks to determine the strength of a component or structure without physically breaking or damaging it, so that it is still usable following the test.

Figure 3.14 Albert Thurston (centre) in front of his Avro 529, 1917.

have perished during the conflict were it not for Thurston's calculations. Figure 3.15 offers a diagrammatic explanation of how an undercarriage stress test was conducted. The aircraft was inverted, and the struts forming the box section to which the landing gear struts attach were strengthened. Weights were then placed symmetrically on a plank suspended from the wheel axle until structural failure occurred.

He was primarily using fundamental mathematical equations and principles to determine the load values at which various components would capitulate, and he was certainly an early ambassador of the 'fail-safe' ethos. Many of the stress loading tests on wings were actually performed at Farnborough. Sand bags acted as weights to be placed on the inverted wings, and a pulley system was used to attempt to raise the aircraft fuselage, thus simulating an aerodynamic lifting force. The result of an actual destructive test on the main wing spars of a Sopwith aircraft is shown in figure 3.16. A number of early flying accidents in biplanes were actually caused by wing struts collapsing during diving manoeuvres. It was only subsequently realized, rather counterintuitively, that

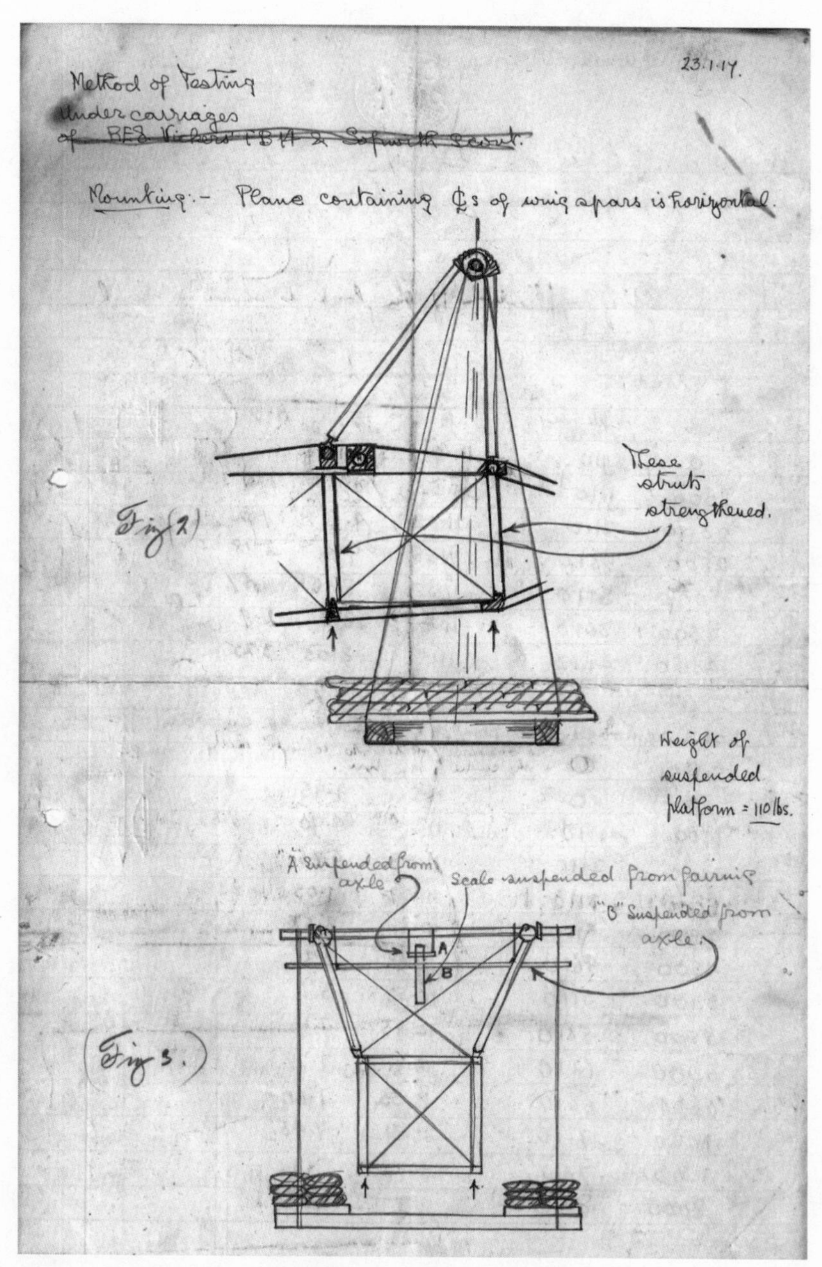

Figure 3.15 Notes on a destructive test on undercarriage struts.

Figure 3.16 Destructive test of a Sopwith SS at the Royal Aircraft Factory in June 1914.

the upper wing could produce a crushing force in such circumstances that was able to induce this type of failure, and stressing calculations were modified accordingly to compensate.

Figure 3.17 shows an extract from Thurston's personal notes in which he performs a simple calculation to determine the load under which a complete failure of the structure will occur. He first computes the total weight of the engine, coolant, propeller, and sundries of the D.H.4 under test as 1,089 pounds. Load, in the form of bags of shot, is then incrementally placed on a platform (weighing 350 pounds) on top of the engine, and he observes any structural issues. Some buckling is apparent when the total load reaches 8,050 pounds. This corresponds to an in-flight load factor of 8,050/1,089 = 7.4; in other words, some damage will be caused to the engine mounting if the pilot of this type of aircraft pulls 7.4g in flight. Shot continues to be added and Thurston notes any subsequent failures. Sudden and complete failure occurs at a total load of 10,150 pounds, or a load factor of 9.32. A comprehensive collection of original notes detailing many of the destructive tests conducted by Thurston and his team based at the ELC can be found in the John Turner MacGregor-Morris archive at QMUL.

Given his practical apprenticeship with engineer and inventor Hiram Maxim, subsequent academic consolidation, and the contribution he made during the war and beyond, one should view

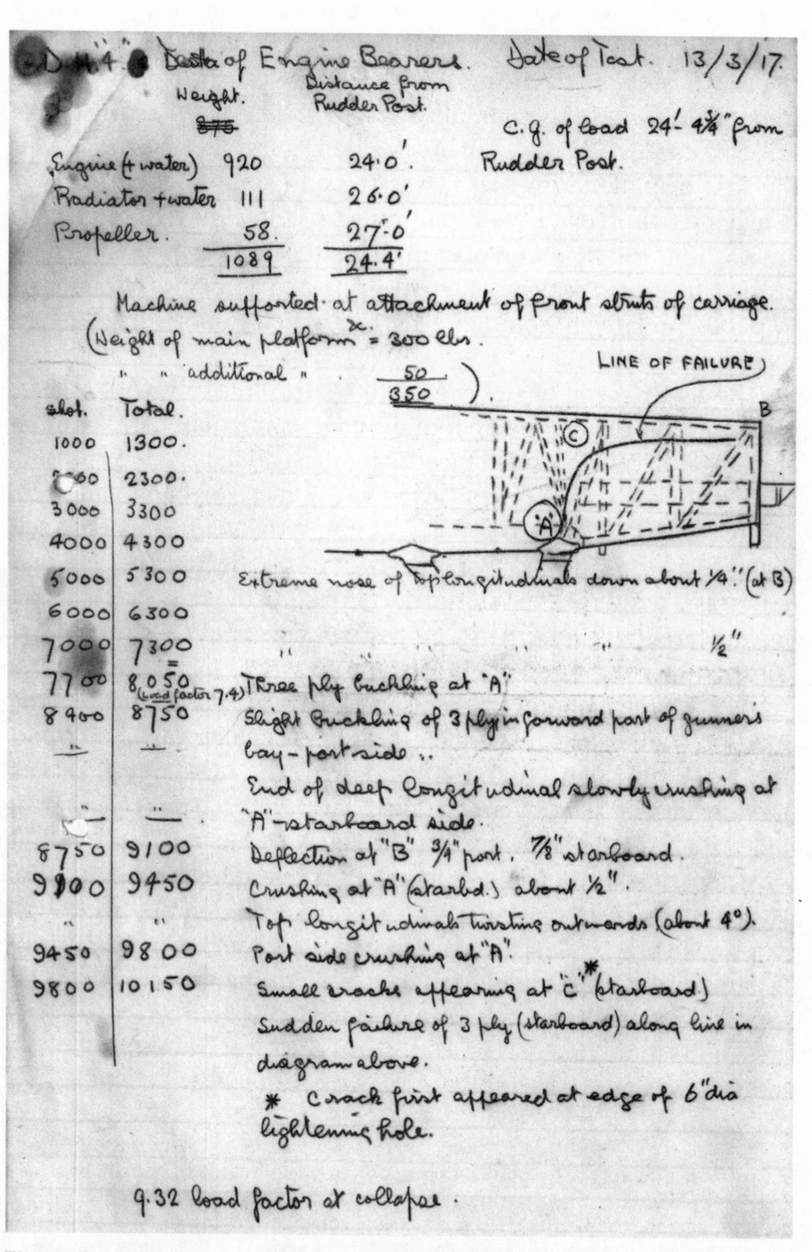
"D.H.4" Tests of Engine Bearers. Date of Test. 13/3/17.

	Weight.	Distance from Rudder Post.
	~~875~~	
Engine (+ water)	920	24.0'
Radiator + water	111	26.0'
Propeller.	58.	27.0'
	1089	24.4'

C.G. of load 24'-4¾" from Rudder Post.

Machine supported at attachment of front struts of carriage.

(Weight of main platform x = 300 lbs.
" " additional " . 50
350).

LINE OF FAILURE
B
C
A

slot.	Total.	
1000	1300.	
2000	2300.	
3000	3300	
4000	4300	
5000	5300	Extreme nose of top longitudinals down about 1/4" (at B)
6000	6300	
7000	7300	" " " " 1/2"
7700	8050 (Load factor 7.4)	Three ply buckling at "A"
8400	8750	Slight buckling of 3 ply in forward part of gunner's bay - port side.
"	"	End of deep longitudinal slowly crushing at "A" - starboard side.
8750	9100	Deflection at "B" 3/4" port. 7/8" starboard.
9100	9450	Crushing at "A" (starbd.) about 1/2".
"	"	Top longitudinals twisting outwards (about 4°).
9450	9800	Port side crushing at "A".
9800	10150	Small cracks appearing at "C"* (starboard)
		Sudden failure of 3 ply (starboard) along line in diagram above.

* Crack first appeared at edge of 6" dia lightening hole.

9.32 load factor at collapse.

Figure 3.17 Destructive test on engine load bearers on a D.H.4 aircraft.

Thurston as one of the true British aviation pioneers. But, as with Dunne, Thurston's name does not feature as centrally as it might in the annals of British aeronautical history, though the material in the QMUL archive makes perfectly clear the crucial role he played in helping to make the pursuit of aviation safe, particularly in the military context. As an academic aside, Thurston's doctorate was the first of its kind, in that it was seated firmly in aeronautics.[21] It helps us appreciate the broad range of interests Thurston pursued within the field and also gives us a sense of the willingness of the editors of the general engineering and aeronautical journals to publish relatively complex technical articles on aeronautics for their readerships.

The first decade of the twentieth century thus drew to a close with solutions having being found to overcome many of the initial challenges aeronautics had faced. America had led the way in 1903, but the efforts of the likes of Dunne had allowed Britain to make up for lost time. The first aeronautical industrialists had established themselves in business, flying schools were opening their hangar doors, and a rudimentary network connecting those in the field had been created through private correspondence and via the contemporary literature, particularly *The Aeronautical Journal* and *Aeronautics*. The British government and the research establishments and universities had responded to the new challenges and opportunities aviation posed and created, with the ACA as the kernel. *Stability in Aviation* perhaps marked a watershed. Powered aircraft had become a regular feature of the skies over Britain by 1911 but were generally unwieldy, temperamental, and unreliable machines. Bryan's mathematics went some way towards defining what was required in practical terms to move beyond these rudimentary vehicles to more sophisticated derivations. To promote more advanced applications of air power, aircraft needed flying characteristics that gave pilots more time to think, more time to look around to observe and navigate, and more time to perform other cockpit duties. An

[21] Thurston's doctorate by publication comprised the following papers: 'Air friction' (unpublished), 'The distribution of pressure on inclined aerocurves' (*Engineering*, dated 20 September 1912), 'Whirling table at East London College' (*Engineering*, dated 8 November 1912), 'The stability of aeroplanes' (*Engineering*, dated 19 May 1911), 'Report presented to the Laboratory Committee of the Aeronautical Society on the wind tunnel erected at East London College' (*Aeronautical Journal*, dated April 1911), and 'The aerodynamic resistance of bars, struts, and wires I' (*Aeronautical Journal*, dated April 1911).

inherently stable aircraft would, in other words, allow a pilot to free up the additional mental capacity required to actually operate the aircraft rather than just control it. The stage was set for Ted Busk as the orchestra played the opening bars of the overture to war.

4

Read All About It!

THE EMERGENCE OF AERONAUTICAL LITERATURE

Britain saw a proliferation of aeronautical journals and magazines between 1909 and the outbreak of World War I. Cody's success, the emergence of the aeronautical industrialists, and the more serious stance on aeronautics being taken by the government meant the time was ripe for more concerted media involvement.

Potential profit was doubtless a motivator, but there was more to this expansion of the written aeronautical word than just money. Those involved came at the task of reporting on aviation matters from different backgrounds and with diverse agendas. Each journal adopted its own catchy slogan to woo its readership, and, as in any competing media market, there was a scramble to secure sponsorship and come up with as many exclusive stories as possible. The sponsors tended to be manufacturers of aircraft or aircraft parts and paraphernalia, or the various flight schools that were emerging all over the country, touting for trade.

Some of the technical information included between the covers of these publications was quite staggering. The level of detail in the technical drawings and in the exposition of mathematics that often supported the articles and discussions was an impressive facet of the literature, and it demanded the presence of technically minded individuals on the payroll. In many respects, these 'technical men' (and they were all men) were more responsible for making the mathematics of aeronautics known to the general public than any others. They were not necessarily highly qualified in mathematics or engineering, but they certainly knew enough to facilitate those who were – technical editors in the truest sense.

Figure 4.1 Stanley Spooner.

FLIGHT

The first, and arguably the most popular, aeronautical publication of the 1909 contenders was *Flight*: a weekly magazine edited by Stanley Spooner that was the official organ of the Aero Club of the United Kingdom. Spooner was educated at King's College, London, and he came into aeronautics from the world of motor cars, having founded *The Automotor Journal and Horseless Vehicle* in 1896. He was a close friend of the Wright brothers – a liaison that prompted aeronautical articles to be carried in his journal – and by 1908 Spooner decided that a dedicated publication was needed: hence the first appearance of *Flight* on 2 January 1909. The magazine was also fortunate to have Algernon Berriman, a talented technical editor, on board.

It was not long before mathematics started appearing in *Flight* articles. A typical example is the series of contributions from American military engineer George Squier, the first of which was seen on 27 February 1909 (Squier 1909, 121). Squier, a major in the US Army's Signal Corps, held a PhD from Johns Hopkins University, Baltimore, and he made a good attempt at generalizing many of the known design and operational concepts in aviation in 1909 using rudimentary mathematics to illuminate his paper. One of the topics he discusses is the mathematics of buoyancy relating to dirigibles, which was simply a play on Archimedes, but Squier soon turned his attention to machines for which 'buoyancy is practically insignificant' (fixed-wing aircraft) and talked about the 'dynamic reaction of the atmosphere itself', drawing upon Duchemin's equation

Figure 4.2 George Squier.

modified for small angles of attack:[1]

$$P = 2k\sigma AV^2 \sin(\alpha),$$

from which, assuming P and α are kept constant,

$$AV^2 = \text{constant}.$$

Applying this to the Wright machine, he then speculated about the potential for faster machines that would require smaller wing areas before reaching the crux of his analysis: the effect of what he termed arched surfaces (today's aerofoils). Using data acquired by his countryman Samuel P. Langley, he came to the startling conclusion that an arched surface was 'dynamically equivalent to a plane surface of 25% greater area than the projected plane'.

For the article's readers, Squier displayed this relationship on an easily decipherable graph, and in fact, graphs, often displaying multiple variables and scales in one picture, abound. They were a common trait of the era. He further derived that the total resistance of the aircraft to the air varies approximately as the square of the relative speed, and the propulsive power varies approximately as this speed cubed – approximations that are still useful today.

[1] Here, P is the normal pressure on a single flat surface, k is a constant that depends on the shape and aspect of the surface (to be determined by experiment), σ is the air density, A is the area of the surface, α is the angle of attack, and V is the relative velocity of translation of the surface through the air.

In addition to considering the drag forces, Squier touched on the importance of stability and control, noting that these were areas yet to be mastered, although the fact that the Wrights had given impressive demonstrations of their flying machine in both Europe and the United States in 1908 was a clear indication that progress was being made in these challenging areas. He makes an illuminating observation regarding the challenges facing a pilot versus those of a chauffeur:

> The fundamental difference between operating the aeroplane and the automobile, is that the former is travelling along an aerial highway which has manifold humps and ridges, eddies and gusts, and since the air is invisible he cannot see these irregularities and inequalities of his path, and consequently cannot provide for them until he has actually encountered them. He must feel the road since he cannot see it.

Finally, Squier's appreciation of the link between the mathematics of hydrodynamics and that of aerodynamics is impressive. He drew heavily upon German physicist and physician Hermann von Helmholtz's insight that[2]

> although the differential equations of hydro-mechanics may be an exact expression of the laws controlling the motions of fluids, still it is only for relatively few and simple experimental cases that we can obtain integrals appropriate to the given conditions, particularly if the cases involve viscosity and surfaces of discontinuity. Hence, in dealing practically with the motion of fluids, we must depend upon experiment almost entirely, often being able to predict very little from theory, and that usually with uncertainty.

It is interesting that Helmholtz himself attempted to compute the various parameters applicable to a potential 'aerial craft' using observed (experimental) data from tests on marine craft. His words demonstrate that the practical limitations of the groundbreaking work of Navier and Stokes earlier in the 1800s were clearly appreciated by those working on fluid dynamics later that century. Even today, computational fluid dynamics and finite-element analysis

[2]Squier uses Helmholtz's paper of 1873 (translated in 1891), 'On a theorem relative to movements that are geometrically similar in fluid bodies, together with an application to the problem of steering balloons', to provide his inspiration (von Helmholtz 1891). As an aside, this is the same paper in which Helmholtz proves, once and for all, that the notion of a human attempting to fly using flapping wings driven by the motive power of the arms is a non-starter.

have to assume certain properties to derive approximate solutions – although, understandably, they get closer to the truth than Helmholtz's pen and paper computations.[3]

This article of Squier's alone, then, illustrates the remarkable way mathematics and principles of aerodynamics were being brought to the general public in real time and in a manner that was comprehensible to many. We should also note Squier's status in the United States. He was a prime mover in the formation of the Aeronautical Division of the US Signal Corps: the equivalent of the Royal Flying Corps (RFC) in the British Army. Just as the RFC became the Royal Air Force in Britain in 1918, so the Aeronautical Division of the Signal Corps eventually became the United States Air Force in 1947. Here was a senior military man in the field of US Army aviation writing mathematical articles for a British journal with the insight of working intimately with the Wright brothers: a real coup for Spooner.[4]

Beyond articles specifically written for the magazine, a contribution from the technical editor would often stimulate a response from readers, some of whom were credible mathematicians or engineers in their own right; indeed, a contribution might sometimes be received that outshone the original article! Joseph Hume Hume-Rothery's multiple offerings proved to be classic cases in point.

A student at Owens College between 1881 and 1886, Hume-Rothery was awarded a BSc (London) in physics in 1887 and spent the next four years at Trinity College, Cambridge. Remarkably, he achieved all this without having attended school prior to college: he was exclusively home tutored. He was called to the bar (Lincoln's Inn) in 1893 and practised law in London for a decade, until in 1904 he became an extension lecturer for Oxford and Cambridge universities. While he might thus be considered as someone with a distant background in physics, he had largely focused on mathematics at Cambridge and was clearly rather good at it, bracketed equal 12th

[3]Today, a computational tool for dealing with the partial differential equations thrown up by problems in fluid dynamics has been developed that finds solutions using the finite-element method: it is named Elmer, in homage to the flying monk Eilmer, Elmer being the modern-day variant of his name.

[4]Although Squier held a senior military rank in the realm of military aviation at the time, he was relatively junior to those wielding true power in the context of the entire US military – a disparity in influence that would undoubtedly have hampered his efforts to execute his mission to promote and advance US aeronautics.

on the 1890 Wranglers' list.[5] He later drifted into the legal profession but kept connections with academia and held a keen interest in aviation.

Hume-Rothery's first published article, 'The X-constant' (Hume-Rothery 1912), questioned the validity of a calculation shown in a previous issue of *Flight* to work out the efficiency of an aircraft, and his article in turn provoked a response. The ensuing debate was an enlightening exchange that does credit to both the author and the technical editor Berriman, who, rather than take umbrage, embraces Hume-Rothery's contribution as a possible step forward for aircraft designers. Berriman had used dimensional analysis and known aerodynamic laws to reach his conclusion, whereas Hume-Rothery employed a more novel approach using 'aerodynamic premises' to arrive at his end point. Within a single page of *Flight*, therefore, we see a technical editor suggesting to those in industry an advancement in aeronautical design theory based on a piece of mathematics put forward by a member of the public.

So impressed was editor Spooner that the readership soon saw more of Hume-Rothery's material, this time looking at negative wing tips (Hume-Rothery 1913a). Spooner's respect for his new-found, amateur contributor was obvious: he declared that he 'treats the question with the proper regard to the fine mesh of the mathematical net that one expects from a Cambridge Wrangler' (Spooner 1913). We are now moving into a realm that is far more mathematically exacting. Not only does integral calculus appear in Hume-Rothery's contributions, but it is clear that the amateur author has read and understood Bryan's work on aircraft stability – no mean feat in itself. But what was the purpose or intent of this article? Bryan's work was undecipherable to most, so any comment on it would have been lost on the average reader. Perhaps a hint is in the foreword by Spooner, in which he talks about Hume-Rothery's line of thought serving as 'a direct link with Prof. Bryan's mathematical treatise' (Hume-Rothery 1913a), the synopsis of which was soon due to be published in the magazine. *Flight* was being used as a means of communication between those intimately involved in aeronautical engineering, in some ways like a distant precursor to group e-mail.

[5] This was the same year in which Philippa Garrett Fawcett became the first woman to obtain the top score in the Mathematical Tripos examinations.

A third article by Hume-Rothery a year on from his first would tackle the prickly subject of the 'vol piqué': the dive that ensues from a stall (Hume-Rothery 1913b). The article – published on 20 September 1913 and continued in the following week's issue – is introduced by Spooner himself, who makes it clear that he had approached Hume-Rothery during the previous Olympia Aero Show specifically to ask him to carry out this investigation.[6] Hume-Rothery states, quite correctly, that 'the question as to the least height in which it is possible to recover horizontal flight after being stalled is a matter of first-class importance to pilots', given the loss of life and airframes due to this phenomenon.

One of the article's diagrams is shown in figure 4.3. We see a multiscaled presentation that depicts all the relevant relationships. Perhaps one driver of this form of display in this situation was copy space, but it is particularly helpful because it clearly shows the evolution of a stall in a way that allows the interrelations between key parameters to be appreciated, although this may only be obvious to someone who has experienced the event. The information needs to be read from right to left. The aircraft is flying at very low speed, and the wings have just lost virtually all their lift. As the aircraft drops in the stall, the angle of attack rapidly reduces and the velocity quickly increases. The wing lift experiences a step change from nothing to a magnitude that will support the weight of the aircraft, at which point it resumes flying.

In his introductory paragraph, Hume-Rothery makes reference to a specific page in the 'Technical Report of the ACA, 1911–12', making it clear that such publications were available to the general public and were being read by them in this pre-war period. He extracts specific data from the report pertinent to the B.E.2 aircraft, and then involves the reader in some quite detailed mathematics, employing Newton's laws of motion, the calculus of variations, and numerous graphs, among other things.

The technical nature of the mathematics shown here is significant. This was by no means the first mathematical article to appear in *Flight*, but it certainly extended, yet again, the boundaries of what was deemed acceptable, mathematically, to present to a general readership. One wonders how many readers would have found this

[6]The Olympia Aero Show first appeared on the aviation calendar in March 1909, featuring a modest display of 11 aircraft in its first incarnation.

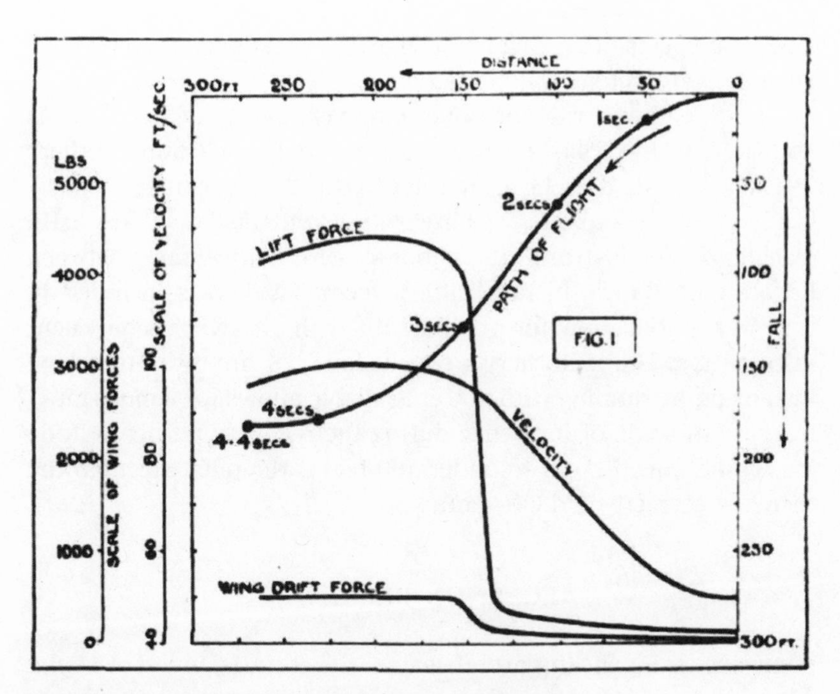

Figure 4.3 Hume-Rothery's depiction of the stall.

level of mathematics comprehensible: very few, presumably. It is not unreasonable to suggest, therefore, that public journals such as *Flight* served a dual purpose, acting as a conduit for semi-scholarly debate among their small percentage of 'expert' readers while also providing information for the general readership.

But if the mathematics itself in Hume-Rothery's article is beyond most, the deductions arising from it are not. A key point here is that what comes out of this analysis has operational, rather than design, implications: it is effectively a member of the public informing 'standard operating procedures' for the RFC and for general aviation. The tentative conclusions were that to recover from an aerodynamic stall with no engine running, a harsh elevator movement to induce a steep dive followed by a gradual increase in the angle of incidence was the best technique to employ to minimize height loss; this would cost the pilot approximately 150 feet of altitude (see the 'Fall' scale versus the 'Path of flight' plot in figure 4.3). The minimum safe flying height was therefore considered to be 200 feet. Of course, this recommendation is rather optimistic since it neglects other factors such as pilot response time and any adverse meteorological conditions.

Hume-Rothery was not done, however; part 2 of his analysis appeared in the following week's issue of *Flight* (Hume-Rothery 1913c). Not only does he again refer to the ACA 'Technical Report', but he also refers to Gustave Eiffel's *Resistance de l'Air* (Eiffel 1910) – evidence of the strong contemporary aeronautical links between Britain and France. In fact, Eiffel's technical data is included in the ACA publication for comparison with the NPL's equivalent information. Hume-Rothery's conclusion this time was that having an engine running with power available allows for a more rapid increase in angle of incidence during the recovery phase; altitude loss could now be kept to under 100 feet if the pilot executed the recovery correctly and efficiently.

ALGERNON BERRIMAN

While occupying his important seat on the editorial board at *Flight*, Algernon Berriman was influencing the dissemination of aviation-related technical knowledge in a number of other ways: via books of his own and through the spoken word. A good example of the latter was his participation in the lectures at the Northampton Institute in London,[7] whose course on aeronautical engineering was included in the mechanical engineering department's offerings.

In 1910, the course was delivered by Frederick Handley Page. He was one of the iconic pioneering aeronautical industrialists, so it was probably quite a coup for the institute to have attracted him to lecture. His course was a complete package: laboratory and workshop practice would be taught on a Monday, aeronautical drawing was to be on Thursday evenings, lectures were given on a Friday at 7.15pm, and mathematics applicable to aeronautics was taught after the Friday lecture. In addition, a series of special lectures were scheduled by guest speakers.

Berriman was one such speaker on 16 November 1910, and he elected to talk on 'The mathematics of the aeroplane': a topic that he later rebranded as 'Mathematics of the cambered plane' and used as

[7] The Northampton Institute was founded in 1894 and became The City University in 1966.

technical copy in *Flight* (Berriman 1911, 58). The lecture was sponsored by the Aeronautical Society, linking the controlling authority in British aviation with mathematics in a popular magazine and the dissemination of current knowledge via the spoken word. Berriman clearly had much to say in the lecture, producing a handout of mathematical formulae for the audience that guided them through the topics under consideration (Berriman 1911, 59). While it possibly did not make the most compelling reading, the fact that Berriman was in a position to talk through all of these formulae indicates that he must have been confident in his understanding of the existing fundamentals of aerodynamics.

Berriman is an important and ubiquitous character in this story, and his application to become a member of the Aeronautical Society reveals something of his background. Educated in Durham in North East England at its College of Science, he studied mechanical engineering. He did not attend university, however, so was very much a practical man rather than a thoroughbred academic. He was given a further opportunity to showcase his knowledge of the aerofoil in a presentation at the 1911 gathering of the BAAS – an event upon which he also reported in the pages of *Flight*: further evidence of the intertwined nature of the aeronautical community, the media, and various influential bodies in British science and engineering.

He cemented his status among the aeronautical community the following year, publishing a book in 1912 entitled *Aviation – An Introduction to the Elements of Flight* (Berriman 1912a), which gave a concise summary of the history and technical state of pre-war aeronautics. The later chapters contained all the mathematics; the rest of the book was written from a more historical and general engineering viewpoint and was illustrated with numerous pictures of aircraft. A particularly unusual aspect of the book is the number of worked examples Berriman presents in order to relate the technical information to application. It was a book that in many ways had something for everyone and was a true reflection of the man's strength as an advocate and exponent of practical aeronautical engineering. Another string to his bow was his keen interest in the engineering aspects of motor vehicle design and construction, and how he considered such vehicles should be driven. This prompted a sequel to *Aviation* that he titled *Motoring: An Introduction to the Car and the Art of Driving It* (Berriman 1915).

THE AERO

On the coat-tails of *Flight*, 1909 saw another aviation publication appear: *The Aero*, a 'penny weekly', edited by the outspoken and controversial character Charles Grey. It offered similar content to *Flight* and was augmented by an annual publication, *The Aero Manual*, which was compiled by all the staff from the magazine. (The staff who worked on *The Aero* also worked on the more established *Motor* magazine: a popular motor vehicle periodical of the time produced by the same publishers.)

Grey, like de Havilland, was a product of the Crystal Palace School of Engineering, and upon leaving there he had immediately taken a post as a journalist at *The Autocar*. It was a time when the automobile industry was taking a keen interest in aviation, and it was Grey's coverage of the Paris Aero Show of 1908 that earned him the commission to edit *The Aero*.

Grey explained the genesis of *The Aero* in the introductory chapter of a book he authored in 1909 under the pen-name 'Aero-Amateur': *Flying – The Why and the Wherefore* (Grey 1909). Essentially, the supply of aviation-related articles and the demand for them had increased exponentially over the preceding year, and the editor of *The Autocar* had decided it was no longer feasible to include them all in the parent automotive journal.

Among the simple diagrams and easily digested explanations that summarize the basic principles of flight and construction, Grey addresses the relationship between mathematicians and those constructing the aircraft. In a chapter headed 'Mathematics v. practice', he recounts a story about Stonewall Jackson, who needed to construct a bridge to get his army across a river during the American Civil War. He instructed his chief carpenter to source the materials and got his mathematicians and engineers to draw up the plans (colloquially referred to as 'them picturs [sic]'). To cut a long story short, the carpenter got on with the task of building and completing the bridge but was still waiting for the plans as Jackson's troops and artillery were safely crossing the construction. As Grey puts it in relation to British aeronautics in 1909 (Grey 1909, 38):

> I have the greatest respect for mathematicians (the more so because I do not understand their mathematics), and one can learn quite a lot of really useful facts and points of design from them; but in the meantime a considerable number of persons are flying very practically, and we are 'still waiting for them picturs' of what a flying machine ought to be.

Figure 4.4 Noel Pemberton Billing.

This was a fair observation on the state of British aeronautics in 1909, a field in which the desire to aviate in anything that could fly was much greater than the desire to aviate safely in something backed by rigorous academic design. The book also contains hints of Grey's passion for, and awareness of, the future importance of air power in the context of national security: something that would occupy much of his time over the following decade.

AEROCRAFT

Yet another monthly aeronautical journal, *Aerocraft*, entered the market in 1909, parodying *Flight* by proclaiming itself to be the 'Official organ of the British Colony of Aerocraft'. It was the brainchild of Noel Pemberton Billing, a rather colourful individual. A runaway as a child, he sampled all manner of occupations, from policeman to boxer. His interest in aviation led him to set up an aircraft design and manufacturing company, Pemberton-Billing Limited, in 1913, but he would soon be distracted by the lure of politics.[8] Elected member of parliament for Hertford in 1916, he became an

[8]Pemberton-Billing Limited would eventually become what is now BAE Systems. Initial funding for the company came out of a substantial bet that Pemberton Billing had with Handley Page, the former insisting that he would be able to obtain a pilot's licence within 24 hours of first sitting in an aircraft. He duly met the challenge, becoming the 683rd licensed pilot in Britain. A biography of Pemberton Billing's remarkable life can be found in Stoney (2000).

outspoken advocate for the creation of an air force independent of the British Army and Royal Navy. He was also a critic of Farnborough's dominating role in British aviation, believing that the standard of aircraft being designed and built there was not high enough to match the UK's German counterparts, and he had views on how best to conduct air warfare, publishing his own book on the subject in 1916 (Pemberton Billing 1916).

So, while adding to the number of aeronautical journals on offer to the public, *Aerocraft* also exposed its readers to more political undertones in its editorials. In fact, Grey and Pemberton Billing became thorns in the side of the establishment – in many ways, they were the self-appointed leaders of the political wing of British aeronautics.

THE AEROPLANE

The Aero was the forbear of a new journal called *The Aeroplane*, which started in 1911 under the editorship of Grey, who had by then established his credentials in aviation journalism by working on the former. He employed two primary article writers: W.F. de B. Whittaker, who concentrated on the more general and political aspects of aeronautics; and P.K. Turner, who wrote on the 'technical and scientific side of aviation' (Grey 1912, 123).

Turner wrote with great clarity of thought, and very much as an engineer. His article 'Second thoughts of an idle engineer' was typical. In it he muses over a number of aircraft design and performance issues, particularly the uniformity of strength in construction (Turner 1912):

> Uniformity of strength is a thing that is worshipped by those who profess most branches of engineering; but it is lamentably neglected in aviation, where it is the most important, for obviously the strength of an aeroplane as a whole is no greater than the strength of its weakest portion, and a greater factor of safety in other parts is not only perfectly useless, but even positively harmful, since, owing to the additional strength, they must be of unnecessary weight, and are therefore inefficient.

While the significant contributions to aeronautical progress made by the likes of Turner and Berriman were, and still are, understated, the passing of Grey in 1953 prompted the editor of *Flight*, Maurice Smith, to write a touching piece to mark the occasion, describing 'C.G.', as he was known to friends, as 'the first great aviation journalist' (Smith 1953). Indeed, Grey's funeral would even attract the presence of Lord Tedder, Marshal of the Royal Air Force, who agreed to give an address, such was Grey's status in the context of aviation. This is quite surprising given the nature of his politics, which are made explicit in an insightful piece carried by *Time Magazine*, written upon Grey's retirement as editor of *The Aeroplane* (Anon 1939):

> As airman, journalist and British subject, Charles Grey ('Center of Gravity') Grey is a bird rare as the wingless kiwi. Editor since 1911 of Britain's well-informed trade weekly *The Aeroplane*, he seldom stuck his balding head inside one, [and] when he did, prayed it would 'land slowly and not burn up'. In a publication ostensibly technical, aerophobic Editor Grey devoted whopping columns to his pet political peeves and peevish political pets. He was shrilly pro-Nazi, anti-French, abominated U.S.-made planes, roundly clapperclawed the British Air Ministry for buying them. A colorful penman with spectacular contempt for fact ('What's the good of that when you can invent your facts as you go along?'), führious Editor Grey perennially brewed bumpy weather in European air politics.

Despite his political agenda, Grey's importance to this story, in the mathematical sense, was his general appreciation of the importance of engineering in aviation and his consequent support for his technical editors, who were afforded ample copy space to bring the mathematics of aeronautics to the masses. If he ever allowed the truth to get in the way of a good story, it would always be in the technical domain.

THE ENGINEER

It is appropriate to mention here the response of the general engineering literature that ran alongside the new journals during the immediate pre-war years. A typical aeronautics-related article from this period is shown in figures 4.5 and 4.6, and there are a number of interesting aspects to note. Being pre-war, there is no attempt to hide the salient design features and performance capabilities of the featured aircraft, the B.E.2; indeed, there is nearly enough information on offer to allow someone who was so inclined to construct the aircraft from scratch, even the specific aerofoil wing section. Additionally, the article clearly indicates that strong links have been

536 THE ENGINEER Nov. 22, 1912

gone by, at the island of Purbeck, the cement stones of the Kimmeridge clay furnished the raw materials for cement-making. But to-day this branch of the industry is extinct, as is also that of using the Septaria nodules of Harwich and the Isle of Sheppey for the same purpose.

Those raw materials which are the product of other industries are at the moment not used in this country, with one exception, however. Chance and Hunt's, of Oldbury, are using the waste from the manufacture of alkali, after being freed from most of its sulphur. The chief constituents of a deleterious nature are sulphates of calcium and sodium, iron sulphide, and free sulphur. There is a variety of substances which lend themselves as raw materials, which no doubt as time goes on will be used. The chief are blast-furnace slag, caustic lime mud, sugar works mud, and ammonia soda waste. The harmful ingredients of this last are calcium sulphate, calcium and sodium chlorides.

THE WORK OF THE GOVERNMENT AERONAUTICAL COMMITTEE.

No. II.*

Returning now to the section of the report dealing with experiments on full-sized flying machines, we propose in this article to give an account of the aeroplane known as the BE2. This machine, a biplane, is of particular interest not only on account of its performances during the course of the recent Military Aeroplane Competitions, but because of the fact that its design has been evolved largely from the information made available by the Advisory Committee for Aeronautics and the National Physical Laboratory. That Mr. O'Gorman and his colleagues at the Royal Aircraft Factory deserve considerable credit for the inception of the machine will be gathered from the following brief summary of its capabilities when loaded with a passenger and three hours' fuel and oil :—(1) Increase its slowest speed by over 75 per cent. ; (2) fly on the level as slowly as 42 miles per hour ; (3) fly on the level as fast as 72 miles per hour ; (4) alight at speeds a little less than 40 miles per hour ; and (5) glide at an angle of 1 in 8. Without a passenger it has climbed to 1000ft. above the earth at the rate of 480ft. per minute and to 6000ft. at an average rate of 380ft. per minute. In other words, it can ascend about as quickly as we can walk with comfort on the level. Its engine is not abnormally large, developing 70 horse-power. It uses ordinary lubricating oil and is moderate in the consumption of oil and fuel.

The engravings on page 537 show the design and construction of the machine, and in what follows some notes on its more interesting features are given.

Landing gear.—It was intended to fit the machine with a new form of landing chassis provided with oil buffers and pneumatic absorbers for taking up vertical shocks. This gear has been omitted in the present instance, but has been tried successfully on a later machine, the BE5. In the BE2 the landing gear comprises two ash skids to which is attached by rubber a tube axle carrying a wheel just outside each skid. The wheels are covered in with fabric in order to reduce the head resistance. It appears that there is reason for believing that this covering-in of the wheels has a slight damping down effect on small lateral oscillations. As will be gathered from the elevation, the landing gear is supplemented with a short rear skid swivelling with but mounted separately from the rudder. When running on the ground at low speeds the machine is thus made manageable to a degree greatly surpassing that which is the case on machines relying solely on the air rudder for this purpose. When running with a fair wind over the ground the BE2 can actually be turned in a radius of 30ft. There is about 80 lb. of load on the back skid when the machine is standing on the ground with the propeller stopped and in this position the main wings make an angle of 12 deg. or 13 deg. with the horizontal. The whole landing gear was severely tested and proved itself satisfactory during the course of the military trials.

Rudder.—The area of the rudder is 12 square feet. At 68 miles per hour and at an angle of 20 deg. it exercises a force of 115 lb. The centre of pressure on the rudder is under these circumstances at a distance of 16ft. from the centre of gravity of the machine. If properly "banked" the aeroplane can be turned in a radius of 60 yards.

Tail plane.—The tail plane has an area of 32 square feet and in flight carries an average load of 35 lb.

Struts.—The section of the struts is of the form designated "Baby," and was adopted in accordance with the results of experiments carried out at the National Physical Laboratory—see The Engineer for last week, page 513.

Wings.—The upper wings have an area of 202 square feet and the lower 172 square feet. In section they are represented in Fig. 2. This section has been adopted in accordance with the results of certain experiments on model aeroplane wings conducted by Messrs. Bairstow and Jones at Teddington. We hope to refer to this interesting series of investigations in a future issue. The dihedral angle between the planes represents a rise of 1½ deg. for each wing.

* No. I. appeared November 15th.

Gliding angle.—The phraseology of the report on this subject is a little obscure, but we infer that with the engine shut off the gliding speed in still air is about 40 miles per hour, and the gliding angle 1 in 6.8. If we are right in our interpretation the performance of this machine closely approaches that of the Maurice Farman biplane entered for the Military Competitions.

Lateral control.—The ends of the planes may be warped 7 deg. in either direction. As the usual flying angle is 2 deg. to 3 deg. this means that the down side of the plane has a maximum angle of incidence of 9 deg. to 10 deg., while the upside of the plane has a negative angle of 4 deg. to 5 deg.

Longitudinal control.—The elevator planes have an

Fig. 1—ROYAL AIRCRAFT FACTORY BIPLANE BE2

The officially reported mean gliding angle for the latter machine was 1 in 6.8 and the mean gliding speed 55.5ft. per second.—38 miles per hour. Gliding at 48 miles per hour the BE2 was found to drop about 500ft. per minute. This is equivalent to a mean gliding angle of about 1 in 8.5. When gliding at

area of 25 square feet and are carried at the rear of the tail plane. The moment of inertia of the aeroplane round the centre of gravity is 1300 in gravitational units.

Engine.—This is an eight-cylinder 70 horse-power Renault engine. When throttled down the slowest

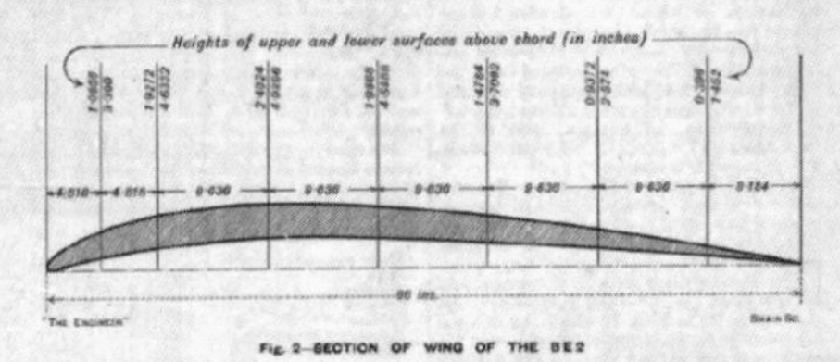

Fig. 2—SECTION OF WING OF THE BE2

66 miles per hour the drop was 1000ft. per minute, so that at this speed the mean gliding angle is 1 in 5.8.

Propeller.—The propeller used is four-bladed and is illustrated in Fig. 3. It has a diameter of 8ft. 6in. and an effective pitch of 6.33ft. The angle of attack is 5 deg. It is designed for a normal speed of 900 revo-

speed of the engine is 200 revolutions per minute. This permits the aeroplane to stand still on the ground with its engine running so that if necessary the pilot can start without assistance. In full flight the speed of the engine can be varied from 1350 to 1950 revolu-

tions per minute corresponding to the acceleration

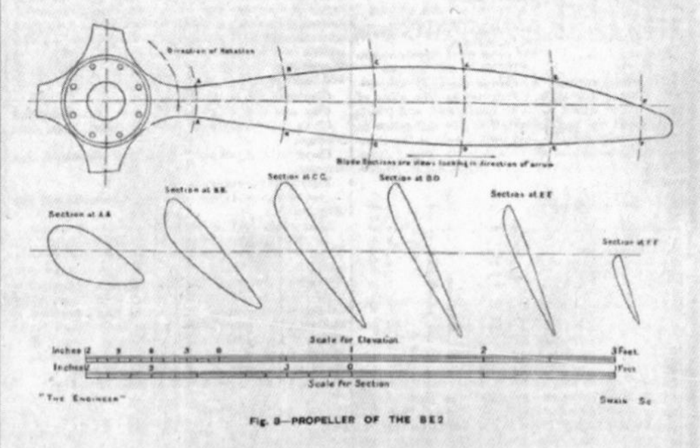

Fig. 3—PROPELLER OF THE BE2

lutions per minute, and a translational speed of 65 miles per hour. The approximate thrust derived from it when driven by a 70 horse-power motor is 296 lb. It is intentionally underpitched so as to spare the engine and give a high horse-power for climbing purposes. A thrust measuring device has been designed to work in conjunction with this propeller.

of the aeroplane from 42 to 70 miles per hour. There is a two-to-one reduction of speed between the engine and the propeller. A silencer has at times been used on the engine. This reduces the horse-power by 2 per cent. under load. From the remarks made in the report we gather that there is more need to silence the valve gear and gear wheels than the exhaust.

Figure 4.5 Article (p. 1) on the B.E.2 from *The Engineer*, November 1912, giving detailed information about its main wing and propeller aerofoil sections.

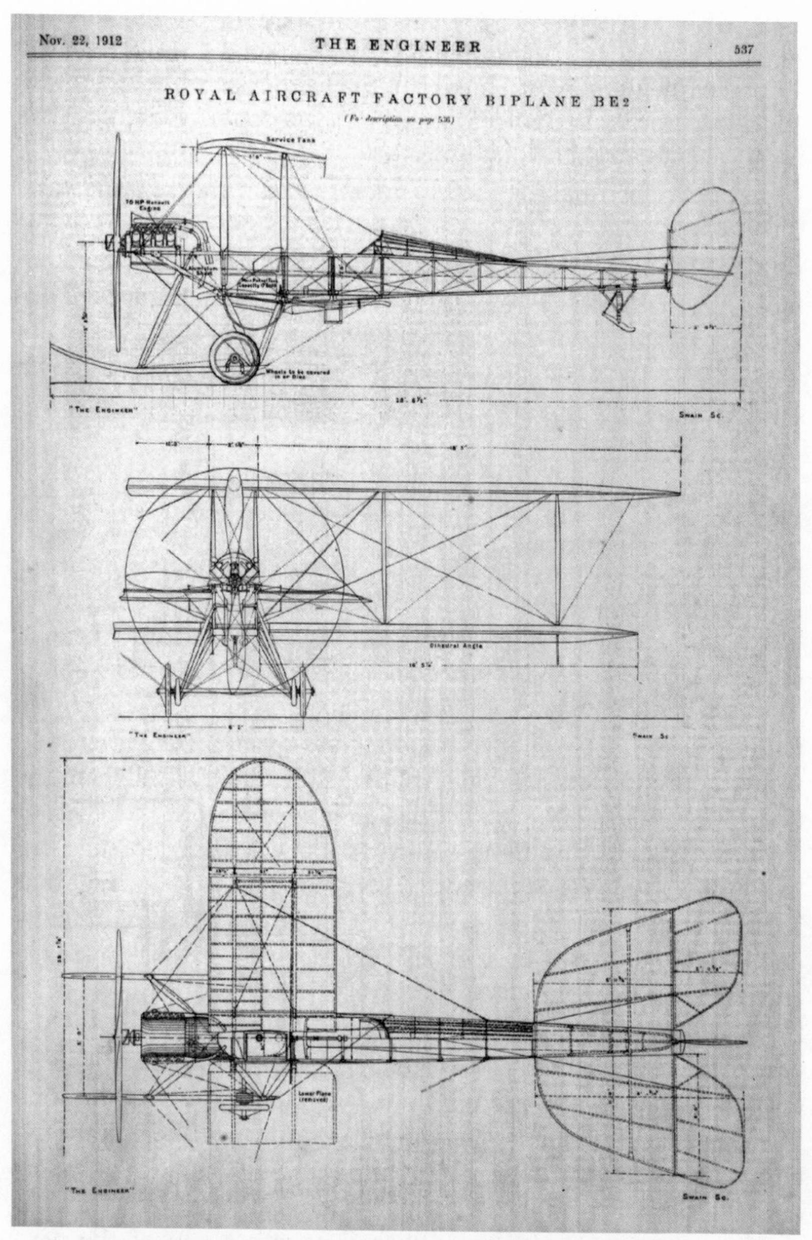
Nov. 22, 1912 THE ENGINEER 537

ROYAL AIRCRAFT FACTORY BIPLANE BE2

Figure 4.6 Article (p. 2) on the B.E.2 from *The Engineer*, November 1912. A detailed schematic showing side, front, and top elevations.

established between the publication's editor and influential scientists and mathematicians working at the research establishments at Farnborough and Teddington: O'Gorman, Bairstow, and Melvill Jones are mentioned by name, with the former being credited with supplying the photograph of the aircraft. The report goes into detail about the structural testing procedures being used at the time, noting that the potential for sand to get into joints and working parts during such tests was undesirable, and that bags of shot were to be used in future. The main mathematical input to the article comes in the form of a power and resistance graph, with a number of different scales and parameters being employed on a single plot to display multiple data, as for the graphs used by Hume-Rothery in his articles.

THE TIMES ENGINEERING SUPPLEMENT

Perhaps to circumvent any possible controversy, there is evidence to suggest that members of the establishment may have resorted to subtle means to bring information to the attention of the general public. *The Times Engineering Supplement* (TES) carried a mixture of technical and general articles on aeronautics, often written under pseudonyms such as 'Pegasus' and 'Ornis'. The nature and content of the reports hint at a source in the upper echelons of the Royal Aircraft Factory. A rather useful and condensed collection of all these articles in the form of original newspaper clippings can be found in two personal scrapbooks belonging to O'Gorman that are held at the National Aerospace Library (NAL) in Farnborough. In one such article, dated 17 April 1912, Pegasus makes an appearance and offers views on the significant stresses that may be experienced by an undercarriage during landing, highlighting the need to take careful account of the wind when calculating an aircraft's kinetic energy. In another article from 1916, Ornis lauds the contributions of Lanchester, Bryan, Busk, and Bairstow in the field of aircraft stability, discussing the impact their achievements had on the ability of the RFC pilots to conduct the operations of war.

These two articles were published at either end of the next important phase of our narrative. Britain had begun to put in place all of the elements required for the British aviation industry to evolve from its disparate, bespoke, and exclusive character into one where the forces and influences were more connected and organized. The formation of the ACA was the catalyst. It was this committee that linked research with resources and aspirations, drawing together the research facilities, the universities, the media, the military, the industrialists, and the government.

It was a period in which certain individuals came to prominence in the field: people such as O'Gorman, Bairstow, and Thurston, who would all play significant roles as Europe marched towards conflict.

Technical information that was often mathematical in nature had started to be catalogued and disseminated. Also, there was a growing influence from the editors and technical editors of the burgeoning number of aeronautical journals that were entering the market, aided and abetted by members of the general public who were able to engage with, and contribute to, certain aeronautical advancements. The scene was thus set for Ted Busk – and those other mathematicians, engineers, and scientists who would follow him at the Royal Aircraft Factory – to prepare Britain for war in the air and see the conflict through to its conclusion.

5

Talisman

EDWARD TESHMAKER BUSK

The Great War began in the summer of 1914, and Britain's airmen took to the skies in aircraft that owed much of their design and performance characteristics to the efforts of one man: Edward Busk. His arrival at the Royal Aircraft Factory in 1912, and the work he did there, significantly affected the trajectory of British aeronautics.

Busk was very much a man on a mission when he first walked through the gates of the Farnborough facility: he intended to build a stable and reliable aircraft for the RFC, and to build it quickly. Given the importance of his task and the fact that he did indeed complete it in short order, it is somewhat surprising that he is not legendary in the world of aviation. His rather distinguished and unique name came from the marriage of his great grandfather, also called Edward, to a Miss Teshmaker: a union that brought a home in Winchmore Hill, London, into the family as well as, at the same time, a son, Thomas. Edward Junior was born in Winchmore Hill on 8 March 1886 to Thomas and his wife, Mary. Edward soon had a sister, Dora, before the family moved to Hermongers in Sussex in 1889, where Ted (as his mother called him) began to show early interest in things mechanical, particularly trains. Another son, Harry, arrived in 1890, and Hans, who completed the family, followed in 1894, only for the siblings to immediately suffer the loss of their father.

Despite the brevity of their time together, it is clear that Thomas had a great influence on Ted's passion for engineering. Mary's decision to move back to London following her husband's death allowed Ted easier access to the engineering museums in the capital, further fostering his growing thirst for technical knowledge and mathematical know-how.

Figure 5.1 Edward Busk with his younger brother Hans.

Ted was originally due to attend Winchester College but, given his bent towards mathematics and science, was instead sent to Harrow. Frustratingly for Busk and his mother, once there, the Master decided that Busk should not specialize in these subjects, but instead gain a more general education. When he reached 18 years of age in 1904, he was sent up to King's College, Cambridge, with a view to taking the Mechanical Sciences Tripos, but he was woefully underprepared due to his lack of specialization at Harrow. He soon caught up, however, and passed top of the second class in his first set of examinations – a feat he repeated in his second year. A final push saw him graduate in the first class in 1907.

During his time at Cambridge, Busk's senior mentor was Bertram Hopkinson, who had been appointed professor of mechanism and applied mechanics at Cambridge in 1903 and was responsible for developing the Mechanical Sciences Tripos course initiated by his predecessor, J.A. Ewing. Hopkinson had a background in mathematics (Trinity College, Cambridge) and experimental physics (King's College, London), and an eclectic range of research interests. These would equip him perfectly for the key role he would play during the war, which culminated in his appointment in 1917 as officer in charge of experimental work at both the Royal Aircraft Factory and the Naval Aircraft Experimental Station at Grain.[1] Hopkinson's life

[1]The Isle of Grain is an area of flatlands and marshes situated on the southern side of the River Thames estuary in the English county of Kent.

Figure 5.2 Busk at Cambridge.

ended on 26 August 1918 when the Bristol Fighter he was piloting crashed in poor weather conditions near London. Reports suggest he became trapped above a layer of low cloud and, in attempting to descend through it, lost control of the aircraft. Being in such close proximity to the ground, he likely had insufficient time to recover the situation once he regained visual references.[2]

Busk's other main academic influences at Cambridge were Charles Inglis (who later became a major) and William Herrick Macaulay, his college tutor in mechanical sciences who would eventually succeed Hopkinson at Cambridge.[3] Inglis and Macaulay, along with Hopkinson, agreed that Busk should remain at the university for a fourth year to continue his studies. Busk was able to do his own research on an engine given to him by Inglis, in a room allocated to him by Hopkinson. The research involved trying to produce a self-regulating paraffin carburettor to give a mixture of the correct strength without adjustment of the air inlet. This involved the study of the flow of air and liquids in pipes and jets and, in the words of a classmate, Captain L.B. Turner, 'was a stepping stone ... to the larger subject of aviation' (Busk 1925, 30).

[2]For a more comprehensive account of his life, see engineer and historian Thomas M. Charlton's essay for the Royal Society (Charlton 1974).

[3]W.H. Macaulay is perhaps best known for his book of 1913 that summarized what was then understood about the laws of thermodynamics (Macaulay 1913).

Figure 5.3 Melvill Jones (during World War II).

Busk won the Winbolt Prize in 1908, primarily on the recommendation of Horace Darwin.[4] Horace was the youngest son of naturalist Charles Darwin and, like Busk, was educated at Cambridge. Following an engineering apprenticeship elsewhere, Horace returned to Cambridge to be employed designing scientific instruments for the university laboratories. Seeing potential in the business, he then founded the Cambridge Scientific Instrument Company in 1881: a concern that would fashion bespoke instruments for all manner of applications. The company played an important role during World War I by supplying instruments to the Ministry of Munitions. Busk had been considering work at established industrial concerns such as Messrs Vickers, but Hopkinson cautioned against this trajectory. Busk's mother endorsed Hopkinson's view, encouraging her son to meet with Walter Wilson and Percy Pilcher, both aviation pioneers, to seek their counsel. He eventually elected to take a place at a more local firm, Messrs Halls in Dartford, where he remained employed until 1911.

[4]The John Winbolt Prize was established in 1904 when John Steddy Winbolt's widow, Christiana, endowed the university with £500. The prize was, and still is, awarded to the graduate that produces the most interesting and innovative paper relating to the profession of civil engineering. Winbolt, an alumnus of Trinity College, was 36th Wrangler in 1864 and spent much of his working life as an engineer in the Far East.

The period immediately following this spell in industry was an important one for Busk since it was when he started to consider the effect of wind on the stability of aircraft. He returned home and converted a barn into a workshop in which he started studying strains in wires and researching the nature and cause of gusts of wind. Writing to a friend in 1911, he stated (Busk 1925, 61):[5]

> I am now starting a series of experiments on flying machine designs, and propose to learn to fly. Before going on to commercial work I wish to complete my first successful machine.

Busk was also the president of a climbing and walking club, at which he spent many hours in the company of Melvill Jones. In 1911, Busk became a second lieutenant in the Corps of London Electrical Engineers.

In February of 1912 he attended the ASL School of Flying. Founded in 1909, ASL (Aeronautical Syndicate Limited) was one of the first aeroplane manufacturing companies in Britain; it diversified into flying training in 1910, based at the Hendon aerodrome in northwest London. At that time, flying licences (known as 'aviators' certificates') were issued by the Royal Aero Club.

Busk's graduation from the school at Hendon coincided with the formation of the Central Flying School (CFS) at Upavon, Wiltshire, which then assumed responsibility for military flying training; prior to this, the military had delegated all its pilot training to the civilian sector. Reports from Hendon on Busk's flying progress appeared in *Flight*, with the initial report noting that (Anon 1912):

> The very unsettled weather during the past week has prevented much flying, but Tuesday proved an ideal day for pupils. Mr Busk put in a splendid afternoon's work. He made numerous flights the whole length of the aerodrome, attaining an altitude of 400 feet, his landings in volplane becoming quite expert.[6] Finally, he successfully negotiated a half circuit, flying very well. Considering that Mr Busk is only able to practise once a week, he is really making most rapid progress, for after only four flying days he handles the machine in a masterly fashion.

[5]He devised a number of instruments to capture wind data, one of which, used to measure the wind force on cylinders, eventually made its way to the NPL for use in research.

[6]Volplane refers to a descent in a gliding mode. For an aircraft with an engine or engines, it is a mandatory part of pilot training to practise controlling and landing the aircraft without power, simulating engine failure in flight.

Figure 5.4 Geoffrey de Havilland.

Busk kept his links with Cambridge and, later in 1912, Hopkinson recommended him to O'Gorman at the Royal Aircraft Factory, who offered him a job. Glazebrook at the NPL was also part of the negotiations; all three seemed keen to secure Busk's talent. Prior to moving to the Hampshire airfield, Busk spent a month's apprenticeship at the NPL working with Bairstow on experiments using the wind tunnel. Here, Busk gained insight into scale effect and various research methods. His first job at Farnborough, however, was to draw upon the experience of Geoffrey de Havilland, the incumbent test pilot, to teach him the fundamentals of test flying.

Initial work was on stability and control, but he was soon burdened with other projects, among them the testing and development of construction materials, instrument design, and photographic techniques. Busk's main assistant was Major Robert Hobart Mayo, who helped design many of the instruments needed for the pre-war experiments.

Mayo was a product of Magdalene College, Cambridge, where he 'showed exceptional abilities in mathematics and the mechanical sciences, with an inclination towards aeronautics'.[7] He headed up the experimental department at the Royal Aircraft Factory during Busk's time there, hence his involvement in the stability programme. As an accomplished pilot, he elected to join the RFC at the start of the war, seeing action in France before being brought back to lead the team undertaking flight testing of aircraft at Martlesham Heath in Suffolk. Busk was messed in Arnold House, the precursor to Chudleigh, where he made the acquaintance of Neville Usborne, and the two soon became close friends and confidants. Usborne was a Royal Navy officer involved in the development of the first rigid airship, and he would perish in 1916 when attempting the first air launch of a fixed-wing aircraft from an airship.

It is worth noting here how adventurous some of the projects being undertaken by this first cadre of pilots were when one considers the rather shaky and uncertain state of the science and the aircraft being employed. It is almost as if the need to innovate in a time of war, to gain an edge, usurped common sense among those giving the orders. While aircraft were still breaking up in mid-air, often for reasons that remained unclear, directives were being issued to attempt all manner of hair-brained schemes. Launching an aircraft safely from solid ground remained challenging enough, and yet, in the case of Usborne, a pilot was being asked to launch from another airborne vehicle: a privilege he sadly paid for with his life. Others were being asked to launch off moving ships, and again this was a pursuit that invariably ended badly for the flight crew.

Returning to Busk, it was evident that conquering the challenge of stability was of paramount importance. Senior army figures had

[7]The quote is taken from Mayo's obituary in *Flight* in March 1957 (Anon 1957).

already earmarked fixed-wing aircraft as reconnaissance platforms, and other potential applications were rapidly being proposed. Busk's starting point was Bryan's bi-quadratic equations. Determining values for the required resistance derivatives to enable solutions to these equations, however, would require more than just a flying mathematician. It needed a cache of original, purpose-made equipment.

On 4 May 1913, Busk was at a Royal Society soirée at Burlington House in London demonstrating the operation of his 'ripograph': an event and device worthy of mention in articles carried in many publications, including *The Times* on 8 May, which reported (Dawson 1913):

> The attention that is being paid to aeronautical research was indicated by several exhibits. Mr Mervyn O'Gorman, Superintendent at the Royal Aircraft Factory, showed the so-called 'ripograph', devised by Mr Busk, which records on one photographic strip the pilot's movements in warping and in steering vertically and right and left, together with the speed, inclination, and roll of the machine and the time.

The purpose of the ripograph was to discern the effect of flight control inputs on aircraft speed, pitch, and roll. In the particular aircraft being discussed, control in pitch and roll was effected by physically twisting the wings. 'Wing warping' aircraft, as they were known, provided the pilot with two large levers, one attached to each wing. Pulling or pushing on the levers had the effect of twisting or warping the wings such that the angle of attack of each wing was altered. Twisting both wings in the same sense would cause the aircraft to descend or climb, depending on whether the rotation decreased or increased the angle of attack, respectively, while warping the wings in opposite directions induced roll.

The ripograph could record how much, and in what sense, the pilot had warped each wing, and simultaneously note the impact of these control inputs on the aircraft's roll, pitch, and speed. As was often the case when new instruments were developed for application in aviation, it would be *Flight* that offered the public the deepest insight into the workings of the ripograph and a second instrument, the 'trajectograph'. These were discussed in the article 'Instruments used on experiments in aeroplanes' (Berriman 1914), which was published the day after Busk's premature death.

The ripograph itself was manufactured by Horace Darwin's Cambridge Scientific Instrument Company and retailed for £7.[8] The small-suitcase-sized and somewhat 'Heath Robinson' device weighed 84 pounds and was a marvel of ingenuity: it was able to measure nine different parameters in flight and was sufficiently compact to sit alongside a pilot or observer in the cockpit. The *Flight* article provided the public with an engineering-based account with no mathematical detail. It spelled out the nine quantities being measured by the device, and gave details about how clockwork and electromagnetism combined with pendulums, drums, pencils, and more conventional components to achieve the task.[9] The interpretation of the post-flight data was, undoubtedly, a major undertaking for Busk, in addition to conducting the flight tests.

The trajectograph was a much more modest instrument that predated the ripograph, having been devised and constructed at the Royal Aircraft Factory around the time of Busk's initial arrival. It comprised only the components needed to measure altitude and airspeed. A single-function device, the 'taughtness-meter', is also discussed; this was a gadget designed by O'Gorman for ascertaining stresses in wires, particularly those contributing to aircraft structural integrity. *Engineering* also carried an article about the instruments in its 26 December issue that year. Further sources of information about these instruments and Busk's work with them are the 'Technical Reports' of the ACA.

November 1913 witnessed the production of the R.E.1, the aircraft that would be used to slot in the final pieces of the stability jigsaw. Busk immediately learned to fly it and began flight trials, one of his more extreme tests being to point the machine at the ground and take his hands off the controls. Thankfully, the aircraft lived up to its billing and automatically righted itself – another example of a pilot going above and beyond what was sensible in the circumstances. Busk took this particular experiment to the extreme edge of the envelope. Pointing any aircraft vertically at the ground causes a very rapid acceleration, and there is a potential for any adverse effects to arise quickly and to rapidly magnify beyond the point at

[8]Approximately equivalent to £600 in 2020.

[9]The conventional components in the ripograph were items such as pitot-static sensors to determine aircraft altitude and airspeed, and a clinometer to measure pitch angle.

which anything can be done to rectify the situation. In conducting this particular experiment in such an extreme way, Busk placed his life in serious jeopardy. The aircraft could easily have broken up due to either the build-up of aerodynamic forces in the dive or the g-force induced during the aircraft's automatic recovery to straight and level flight. With nothing restraining the pitch controls, there must also have been a risk that the aircraft would go inverted. And there was no escape for this brave aviator either: reliable ejection seats were a long way in the future, and there was no parachute for Busk if things went awry.

Luckily, Busk's calculations had been sound and the aircraft passed this and many other tests, allowing him to roll out the B.E.2c on 30 May 1914; this was a stable aircraft that would become the workhorse of the RFC during the first year of the war that Britain would enter a couple of months later on 4 August. The main modifications to de Havilland's original design were a triangular insert in front of the tailfin to assist stability in yaw, a non-lifting tailplane and movement of the lower main wing to compensate for the change in lift moments to improve longitudinal stability, increased dihedral on the wings to enhance lateral stability, and a four-bladed propeller on a more powerful, 110-horsepower engine. The new machine had a remarkable ceiling of 11,000 feet.[10]

In Busk, Britain had a talented mathematician and engineer who had learned to fly and become a respected test pilot, and who was then able to meld his academic ability with his flying skills to produce a revolutionary aircraft. The critical point was that Busk had the mathematical comprehension needed to fully understand Bryan's equations and also to appreciate what experiments were required to realize the design features that would comply with Bryan's criteria for stable flight. There is little doubt that without Busk's efforts, British airmen would have been flying in machines with very different handling characteristics at the outbreak of conflict. Ironically, the stability of the B.E.2c would become a burden and disadvantage when the aircraft's role expanded from pure reconnaissance into the realm of aerial combat, where rapid manoeuvrability facilitated by greater instability, particularly in roll, became a desirable performance attribute.

[10]An enlightening account of Busk's role in the development of the B.E.2c can be found in Rood (2011).

Three months later, and with the war well under way, Busk was now in charge of the department at the Factory that was responsible for aerodynamic design and research, the testing of aircraft materials, aerial photography, and aircraft instrument design. The war had also brought with it military hardware, such as bomb-sights, that needed development and testing.

In the late afternoon of 5 November 1914, Busk and de Havilland climbed aboard their respective B.E.2c aircraft, presumably for their final flight tests of the day. The main fuel tank of the B.E.2c was situated behind the pilot in the passenger compartment, and fuel had to be manually pumped forward into a collector tank nearer to the engine. The pilot, therefore, had to periodically reach down to locate and operate a pump handle to move fuel forward as required. The downside of this system was that any leak or overspill was not always immediately apparent. As the two pioneers passed 1,000 feet in the climb, de Havilland recalled seeing a huge flash out of the corner of his eye.

Nobody knows for certain, but it is thought that a stray spark from the engine had ignited fuel that had leaked and pooled in Busk's cockpit. Mary Busk describes the scene as told to her by eyewitnesses (Busk 1925, 85):

> The aeroplane burst into a sheet of flames ... drifted aimlessly for a few moments, and it was evident, therefore, that there was no pilot to guide it, and it glided downwards as far as Laffans Plain, where it fell to the ground. He must have passed into another life so suddenly that one can hardly call it Death.

In an unequivocal and heartfelt statement to Busk's mother, General Sir William Sefton Brancker, Director of Military Aeronautics, said (Busk 1925, 91):

> Your son is an irreparable loss to the British Army and, indeed, to the nation, for there are few men available with a like combination of an exceptional brain and scientific knowledge with perfect courage.

His words were penned in a letter to Mary Busk on 7 November 1914, two days after her son's flying accident.[11] General Henderson

[11]Brancker's letter is one of a number included in a book written by Mary Busk about her sons Edward and Hans, both of whom died in flying-related incidents (Busk 1925). Edward and his father, Tom, are commemorated in stained glass windows in Holy Trinity Church, Rudgwick, a short distance from Hermongers, the family home.

considered Busk's death to be one of the greatest losses aeronautics had ever suffered.

The legacy of Ted Busk was profound – in many ways, he defined the role of an aeronautical engineer. He applied mathematics in a true sense, conducting and reviewing practical experiments that informed design and construction, and he pushed back the boundaries of knowledge in this very new area of engineering. He was also clearly a very brave man. Test flying is a precarious activity at the best of times, but in those early years of fixed-wing flight it must have been extremely challenging. Busk exemplified the perilous practice of flying aircraft while applying mathematics. His loss was a huge blow, but the benefit of having academics working on the problems arising in aeronautics had been proven. Even prior to Busk's demise, O'Gorman had commenced the task of headhunting the would-be members of the elite technical team that would largely be responsible for the rapid progress seen in British aeronautics throughout the war years. The first to arrive at Farnborough, just prior to the tragic loss of Busk, was G.I. Taylor.

THE METEOROLOGIST

The list of outstanding issues in aeronautics that required attention in 1914 was a long one, despite the fact that Busk had by then solved the stability problem. One significant unknown was the precise distribution of air pressure over an aerofoil wing in flight, and how this distribution altered as the aircraft manoeuvred or changed speed, and it was Geoffrey Ingram Taylor who was asked to look into this aspect of aerodynamics.

Cambridge-educated Taylor would become very much the intellectual odd-job man of the group. He worked in 'H-Department', which was a bit of a catch-all facility. Housed in a tin-roofed building next to the airfield, this was where much of the aerodynamic work was done – often, according to Taylor, to the sound of aircraft crashing outside.

H-Department was a multidisciplined unit staffed by a small number of scientists, mathematicians, and engineers, specializing in chemistry, fabrics, instruments, mathematical calculation, and materials testing. It was located very close to the flight line, which facilitated immediate technical intercourse between the ground-based academics and the test pilot cadre. Taylor had been introduced to practical fluid mechanics while working as a meteorologist

Figure 5.5 The proximity of sheds and offices to the test aircraft dispersal at Farnborough.

on a whaling ship in the North Atlantic: a mathematical grounding that would serve him well both at Farnborough and beyond.

On the day war broke out, Taylor immediately offered his services in person at the War Office, volunteering as a meteorologist. The army insisted it had no use for a weatherman, and suggested that he report to the Royal Aircraft Factory instead. This he duly did two days later. The dismissal by the duty officer at the War Office regarding Taylor's offer of help with weather prediction raises a smile: 'Soldiers don't go into battle under umbrellas – they go whether it is raining or not' (Taylor 1966, 109)! The impact that weather has had on various military actions throughout history, and would have on a number of World War I campaigns, makes one wonder if this dismissal of Taylor's meteorological expertise was somewhat short-sighted.

The directive to point mathematicians and scientists towards Farnborough had been given by Brancker, who realized, with some foresight, that he would be able to put to good use as many scientists as he could find. Meteorology's loss was certainly aerodynamics's gain, in any event.

O'Gorman met Taylor soon after his arrival and immediately introduced him to Busk, who suggested Taylor join the mess at Arnold House, then managed by Thomas Cave-Browne-Cave, who had arrived at Farnborough in 1912.[12] At this early stage of Farnborough's expansion, the only major players at the base who were not in the Royal Navy were Taylor, Busk, and industrial engineer and engine specialist Frederick Green. Others who soon joined Taylor to bolster the academic ranks at the Factory were Francis Aston, George Thomson, Frederick Lindemann, William Farren, Henry Fowler, and Edgar Adrian, all of whom will be discussed presently.

Taylor describes, with some humour, one of the first tasks he was given in his new job, which was to design a British equivalent to the French darts that could be dropped on enemy troops from an aircraft (see figure 5.6) (Taylor 1966, 109):

> The French had used pointed 'flechettes' but we found that they whirled when dropped, even though the centre of pressure when falling obliquely was well behind the centre of gravity. In any case it seemed to be a matter of prestige that we must have English darts so we designed and tested some with the necessary aerodynamic properties. Our last test was to find their distribution on the ground when dropped from an aeroplane out of a container, and we got a pilot to drop them onto an unused piece of ground. To measure their distribution we searched the ground and whenever we found the rear end of a dart projecting from it we pushed a square piece of paper over it to mark its position. We had just completed this work, and were preparing to photograph the field when a cavalry officer came up and asked what we were doing. When we explained that the darts that covered the field had been dropped from an aeroplane he looked at them, seeing each piercing a white square and said in a surprised tone, 'If I had not seen this I would never have believed it was possible to make such good shooting from the air!'[13]

The first major technical issue addressed by Taylor was the weakening of propeller shafts when channels were cut in them to route necessary controls. In conjunction with Alan A. Griffith, the pair developed an analogue computer that was able to determine the mathematical solutions for the stresses induced in the square corners of the channels, which were the recognized weak spots.

[12] Cave-Browne-Cave became the British authority on non-rigid airships, and following the war he became a professor of engineering at Southampton University. One of his sisters, Beatrice, worked in the field of aircraft stress analysis during the war: a contribution discussed in chapter 7.

[13] The darts were never deployed because senior army officers regarded them as inhuman: weapons that could not possibly be used by gentlemen.

Figure 5.6 Postcard by E. Le Deley of Paris depicting an attack on a German gun battery by French aircraft using flechettes.

Griffith studied mechanical engineering at the University of Liverpool before his arrival at the Royal Aircraft Factory in 1915 to work with Taylor. The two men pioneered the use of soap films in solving torsion problems, and Griffith became a specialist in the field of stress-induced crack propagation in materials. He would later play an influential part in the development of the jet engine in Britain – in particular, on how such an engine might be used to lift an aircraft vertically off the ground.

As Taylor's remit broadened, he became increasingly frustrated by not being able to get a complete picture of airborne experiments conducted by third parties. He thus concluded that the only efficient way for him to fully understand aerodynamic problems was to train as a pilot himself and conduct any airborne tests that were necessary. He duly circumvented the official opposition to such a notion, took a sabbatical at Brooklands airfield, and learned to fly on a Farman Longhorn (figure 5.8).

Taylor's recollections of this experience are frank, honest, and witty. 'I must confess, however, that I was bad at judging distances,' he laments. 'My first landing was too far along the field and ended in the sewage farm just outside it!' He was certainly a pilot who relied more on his cockpit instruments than most:

Figure 5.7 G.I. Taylor.

> I had complete faith in my instruments and did not look at the ground till it was necessary to land on it. A rumour even went round that on one occasion this faith was misplaced because while I was turning, as I thought at about 2000 feet up, the tail of my eye caught sight of a tree quite close to. It turned out the instrument I was relying on was the rev counter, not the altimeter.

He was also guilty of losing a vital component of what was 1915's state-of-the-art air-to-ground communications equipment (Taylor 1966, 110):

> At that time wireless telephony had not been invented and the transmitting apparatus had a long wire aerial with a weight at the end which the pilot had to let down when he wanted to transmit. I failed at the first attempt because I forgot to wind it up again before I landed and it was torn off near the edge of the airfield.

I find Taylor's nonchalance in the way he recalls these events both endearing and extraordinary. Missing the runway and finishing up in a sewage farm on one's first solo attempt at a landing is hardly a confidence-building start to a flying career, yet he seemed content to just dust himself off (or perhaps dry himself out!) and have another go.

This theme of resilience in the face of trauma is a common one among the flying academics at the Royal Aircraft Factory. The fact that Taylor was more reliant on his cockpit instruments than on visual clues to control his aircraft is also fascinating. Flying instruments in the modern era are generally sophisticated, accurate, and

Figure 5.8 The Farman Longhorn.

reliable, but the same qualities were not guaranteed in 1915. Artificial horizons were only just becoming available, compasses suffered from confusing turning errors, and the accuracy of just about every device mounted in the cockpit was, to some extent, adversely affected by engine vibration.

Following his flying training, Taylor returned to Farnborough to tackle, via a sequence of airborne trials, the prickly subject of pressure distributions over an aerofoil in flight. Busk was no longer alive to work with Taylor, but the former's legacy, the stable B.E.2c aircraft, lived on and was essential to the success of Taylor's work. The point of the trials was to gather data that could be compared with equivalent data recorded from a model of the same wing in Bairstow's wind tunnels at the NPL. Conversion factors were the goal: if aerofoils could be tested in the tunnel and the results scaled up to apply to the full-sized wing, the implications were huge, as much of the often risky work that was necessarily being done in the air could then be transferred to the relatively safe and inexpensive environment of the tunnel.

Before commencing the trials, Taylor first had to design a system that could physically measure the pressure at various points over the wing, recording the results in real time; thus was born Taylor's multiple manometer (Taylor 1966, 111):

> I got the workshops to make a metal strip with 20 pressure holes and insert it in the wing of a B.E.2c machine. This

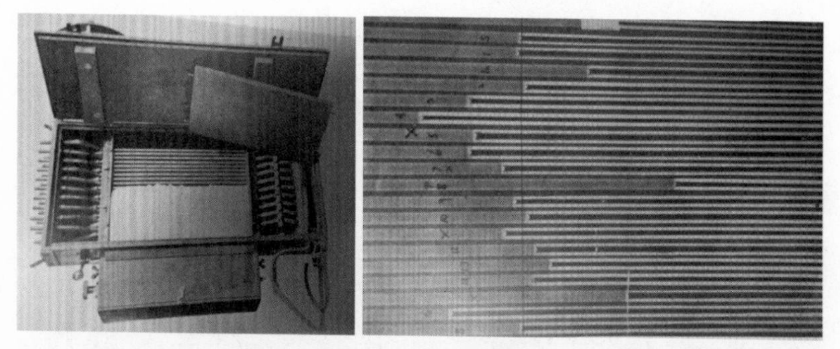

Figure 5.9 Left: the manometer. Right: photographic data.

> was a really stable machine. I could not have made the measurements I did if it had not been stable. I made a multiple manometer with photographic paper which could be wound over the bank of tubes, each of which was connected with one of the pressure holes, and arranged a light which I could switch on for an instant when I had the machine flying level and steadily on course on a calm day.

The manometer itself is shown on the left-hand side of figure 5.9 and a typical record from the photographic paper can be seen on the right-hand side. Once conducted, the trials revealed major discrepancies between wind tunnel predictions and actual measurements taken using the full-scale aircraft. It was subsequently realized that these differences were due partly to the effects of the walls of the tunnel, which modified the general airflow around the aerofoil and, in particular, affected the nature of the wing vortices. Figure 5.10 shows some of Taylor's results. The upper part of the diagram indicates the positions of the holes that Taylor had requested be drilled in the wing section. This is a cross-sectional view of one of the wings of Taylor's B.E.2b aircraft, and it is apparent that interest is most concentrated around the leading edge of the aerofoil. During the trials, Taylor flew the aircraft at speeds between 50 and 97 miles per hour, and the lower portion of the diagram plots the pressure coefficients at each hole over this range of speeds.

The pressure coefficient C_p, which is basically the ratio of pressure forces to inertial forces, is found by dividing the difference between the static pressure measured on the wing, P_m, and the free-stream static pressure, P_{fs}, by the expression for free-stream dynamic pressure, ρV^2:

$$C_p = \frac{P_m - P_{fs}}{\rho V^2}.$$

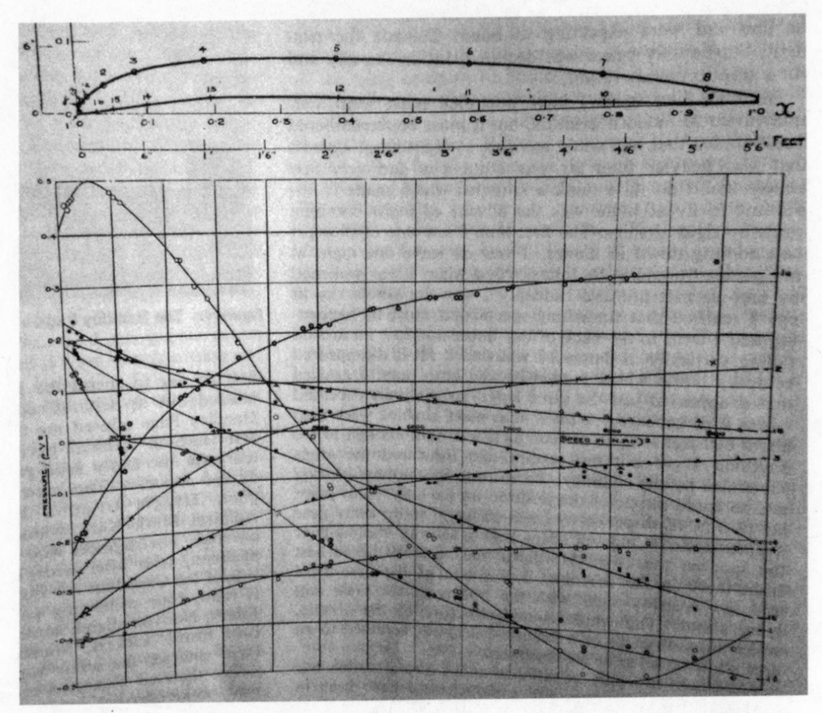

Figure 5.10 G.I. Taylor: pressure distribution over an aerofoil in flight.

Here, ρ is the air density and V is the air velocity, taken as the aircraft speed.

In Taylor's diagram, the ordinate axis is plotting C_{p} over the range ± 0.5. Positive values of C_{p} indicate that the static pressure measured on the wing is higher than the ambient static pressure away from the wing; negative values of C_{p} indicate the opposite.[14] The abscissa plots speed (in miles per hour) squared and ranges from 2,500 to 9,500.

Each numbered curve shown in the lower diagram thus relates to a corresponding hole in the upper diagram. The shape of the curve for each hole shows how, at that particular point on the wing surface, the static pressure in relation to the free-stream static pressure changes as the speed of the aircraft changes. Taylor does not offer any details of how he actually obtained the free-stream value of static pressure required for his calculations, but one assumes it

[14]It is not clear why Taylor omitted the $\frac{1}{2}$ in his expression for dynamic pressure, which is ordinarily calculated as $\frac{1}{2}\rho V^2$.

was via an instrument mounted away from the aerofoil in undisturbed flow. In addition to pressure distribution information, Taylor would have been able to use some sort of integral method to derive an approximate value for the total lift generated by the wing at any given speed.

The drag experienced by an aircraft in flight was also of concern to Taylor. Ludwig Prandtl's advanced mathematical treatment of wing tip vortices had not reached Britain, so Lanchester's more simplistic analysis was all Taylor had to go on in this regard. Furthermore, Taylor admits that he had not thought of considering wing drag as comprising two distinct components: form drag and induced drag (see p. 235 for definitions of these categories of drag).

Having provided an important insight into the characteristics of the aerofoil, Taylor went on to look at the nature of air turbulence and its potential impact on aircraft. He had experienced the effects of these lumps and bumps in the atmosphere when flying the B.E.2b, a variant of the B.E.2 series of aircraft that had the distinguishing feature of wing warping controls, a mechanism he found particularly discomforting. Another project he worked on was describing mathematically the action of parachutes – although, remarkably, it was not standard practice for British aircrew to carry one.

A letter from Taylor to Lindemann illustrates the former's sense of humour and also gives us insight into the relationship between Chudleigh and the popular aeronautical press. Sent to Lindemann prior to his and Farren's return from their flying training, Taylor notes:

> When Farren did not return on the appointed day and when no news of you came to hand I began to get anxious, so I set up a seismograph to register the number and magnitude of your landings. Unfortunately Thomson started mending his bike with a hammer and since then the poor seismograph has been in the hands of a skilled mechanic. I see *The Aeroplane* has been having a go at Farren this time. Grey has apparently published some accurate figures this time so perhaps the little beast may get run in under the Official Secrets Act.

The banter was clearly as harsh between fellow aviators then as it is now, but it is Taylor's contemptuous remark regarding Grey that is most revealing. It tells us that aeronautical journals were being read by those at the Factory and that perhaps not all of the

Figure 5.11 Francis Aston.

information being printed by these publications was as accurate as it might have been.

While we have glimpsed some of Taylor's technical contributions to aeronautics above, what this fails to convey is the character of the man and his strength as a scientist and mathematician. His personal recollections of the World War I era are told with such dry humour and modesty that the danger to which he was exposed during his time at Farnborough is somewhat hidden (Taylor 1966). Later in his career he would confirm his technical credentials by being an important member of the British contingent that formed part of the Manhattan Project to develop the first atomic bomb.

It is perhaps unfortunate that the stellar nature of Taylor's post-World War I work in many ways overshadowed the merit and value of his efforts at the Royal Aircraft Factory. His remarkable and diverse achievements in aeronautics are barely a footnote amid his huge portfolio of more than 200 pieces of published work, now collated in four large tomes (Batchelor 1958–71). Taylor's interests were also very broad: he bridged the gap between applied mathematician, engineering scientist, and classical physicist.

Taylor's academic companions at Farnborough during the early years of the war would also go on to make significant contributions in the world of science and engineering. Francis Aston, for example, was a chemist who had been working at the Cavendish Laboratory in Cambridge prior to his appointment at the Royal Aircraft Factory. After the war he invented the mass spectroscope and used it to isolate stable isotopes of various elements: work that led to the award of a Nobel Prize in 1922.

As is the case with Taylor, little recognition or emphasis is placed on Aston's achievements at the Factory because his contribution to the war effort was completely eclipsed by his subsequent work in isotopic chemistry. If one looks through his 12-page obituary written by radiochemist George de Hevesy for the Royal Society, the following extract is the only reference to Aston's war work (De Hevesy 1948):

> During the first world war Aston served as Technical Assistant at the Royal Aircraft Establishment at Farnborough and was crashed in an experimental aeroplane in 1914, but escaped unhurt. Here he made use of his chemical abilities in research on aeroplane dope and fabric. He also invented the special type of neon tube for short flashes.

This phenomenon – of a lack of full recognition or detail regarding the activities of those scientists and mathematicians at the Royal Aircraft Factory – is perhaps best explained by the expertise of those writing the obituaries and the time frames involved. De Hevesy, for example, was a Nobel laureate in chemistry himself, so his focus when writing about Aston's life was almost bound to be on the latter's work in this field. The fact that Aston's Farnborough interlude came earlier on in his career will also have made it a distant episode when de Hevesy put pen to paper.

Slightly better insight is thus found in another obituary for Aston written for *Nature* by George Thomson (Thomson 1946). Thomson tells us that Aston was specifically working on the use of pigments in the dope that was being used to protect the aircraft canvas materials from the ravages of the weather. Interestingly, Thomson also lets slip that Aston drew on the academic expertise available among those residing in Chudleigh House to discuss and develop his isotopic theory. Most notably, he conspired with physicist Frederick Lindemann to get a grounding in the relationship between quantum theory and the potential existence of isotopes. Lindemann and Aston even published a joint theoretical paper in the *Philosophical Magazine* soon after the war, indicating that work in this field must have been going on alongside the activities related to Factory aeronautics (Lindemann and Aston 1919). Aston would also combine with Thomson himself to produce an article for *Nature* in 1921 that discussed the constitution of the element lithium (Aston and Thomson 1921).

Figure 5.12 George Thomson in 1937.

George Paget Thomson, son of J.J. Thomson,[15] would also receive a Nobel Prize, in 1937, for his work on the wave-like properties of the electron. 'G.P.', as he was known, was in many ways the glue that held the Chudleigh lot together. He was present at the start of the venture in 1914 and saw it through to its end in 1918, witnessing others come and go, or succumb to the perils of aviation. Thomson's contributions at the Factory were significant and prolific, but one gets the sense that they have been underplayed in the historical record – victim, again, to the high-profile nature and status of subsequent work and achievements.

Accounts of Thomson's time at Cambridge as an undergraduate help us to understand the many connections and similarities between activity at that university and happenings at the Factory. In essence, the Chudleigh lot were a subset of people from a 'Cambridge lot'; just as RFC squadrons were detached to France from bases in Britain, so academics were detached to Farnborough from their base in Cambridgeshire. This meant that productive and influential friendships at Cambridge continued to have relevance to, and impact on, aeronautics at the Royal Aircraft Factory.

Thomson, for example, was a very close friend of engineer William Farren, who is discussed in chapter 7. The two had forged

[15] Joseph John Thomson is credited with the discovery and identification of the electron in 1897. He was awarded the Nobel Prize in Physics in 1906.

their relationship at the Perse School as teenagers, collaborating on engineering projects such as the construction of a working model of a submarine. Thomson followed Farren to Trinity College, and in later life the former, discussing his understanding of engineering concepts and principles, admitted that Farren had been his major influence. Thomson spent his first two years at Trinity studying mathematics. He excelled, not least because he had been attending undergraduate lectures prior to going up to the College, and, of course, he was undoubtedly receiving some high-powered private tuition from his father. Thomson switched focus to study physics in his third year.

In 1919, after Thomson and Aston had both returned to Cambridge, they each discovered, using different methods and equipment, that lithium has two isotopes. This implies that the two would almost certainly have been discussing this research alongside their aeronautical activities while at Farnborough. Since both men did great things in the fields of physics and chemistry after World War I, one imagines their aeronautics interlude was more of an interruption than a passion. That said, Thomson's first major project after the war was to write a book that summarized corporate technical knowledge of the subject. *Applied Aerodynamics* (Thomson 1919) was the up-to-date sequel to Chanute's pre-1900 summary. Thomson's book is discussed in more depth in chapter 8.

Thomson, along with Busk, Taylor, and Aston, exemplifies the multidisciplinary nature of the talents and expertise brought to the Royal Aircraft Factory during the war. They all had mathematics at the core of their knowledge, but each with a different overlay, be that a deeper understanding of certain aspects of physics, engineering, or chemistry.

It would be his understanding of atomic physics that led Thomson down the same path as Taylor when another world war reared its ugly head at the end of the 1930s. Thomson was appointed chair of the MAUD Committee that coordinated research into the feasibility of an atomic bomb, and reports generated by this group eventually gave impetus to both the Manhattan Project in America and, following some industrial espionage, the Soviet Union's atomic weapons programme.

The origins of the name MAUD are quite bizarre. Many thought Thomson had suggested it as an acronym for 'military application of uranium detonation', but the truth was that the name had been taken from the text of a telegram from physicist Lise Meitner to a friend following a visit with the Bohr family: 'Met Niels and Margrethe recently both well but unhappy about events please inform Cockroft and Maud Ray Kent.' When committee member and renowned physicist John Cockroft saw the message, he assumed it to be some sort of code, and Maud was thus taken as the designator for the secret British nuclear weapons programme. Only later did the team discover that the original message was not code at all, and that Maud Ray was actually the governess who had taught Bohr's son while he was in England, and she just happened to live in Kent!

The endeavour also reunited Thomson with Henry Tizard, Rector of Imperial College by this time, and also chair of the Committee for the Scientific Study of Air Defence.

If mathematics, physics, and chemistry seem like disciplines that can be applied naturally to many aspects of aeronautics, a less obvious area of science to trawl for help with something such as navigation is, perhaps, the world of physiology. But it is to just this topic that we now turn in order to consider the work of Keith Lucas, Farnborough's compass guru during the war.

6

A Physiologist and a Physicist

THE PHYSIOLOGIST

Born in 1879 in Greenwich, Keith Lucas was the second son of Francis Lucas, a cabling engineer. Mathematics featured on his mother's side, her lineage containing teachers of navigation and nautical astronomy. After attending Rugby School, Keith went up to Cambridge in 1898, graduating with a first class honours degree in Part I of the Natural Sciences Tripos. He went on to be awarded his DSc in 1911. Lucas was already a fellow and lecturer of Trinity College, Cambridge, and an FRS (1913) when the war began. Among his students at Cambridge were A.V. Hill and E.D. Adrian, with whom Lucas conducted research into the physico-chemical aspects of muscle and nerve interaction. Both Hill and Adrian were seconded to help with mathematical aspects of warfare, and each would go on to receive a Nobel Prize following their return to the world of physiology.

Although officially a physiologist, Lucas had a talent for instrumentation. A director of the Cambridge Scientific Instrument Company since 1906, he brought his expertise to Farnborough on 4 September 1914 and immediately began work on designing a new, more accurate, trustworthy, and responsive aircraft compass. Existing compasses suffered from an effect that Lucas identified and termed 'northerly turning error', which made navigation in cloud impossible and navigation in general somewhat of a lottery. He also worked on bomb-sights, designing the first gyro-stabilized system, and developed the 'photokymograph': a piece of equipment that could measure aircraft oscillations using the sun as a reference.

Robert Mayo generally flew the aircraft while Lucas was conducting experiments, but in 1916 Lucas arrived at the same conclusion

Figure 6.1 Keith Lucas: Royal Aircraft Factory ID picture.

as Taylor: he would gain far more from his experimental flights if he was piloting the aircraft himself rather than just observing. He therefore took advantage of an opportunity to move to the Central Flying School at Upavon to commence flying training. However, on 5 October, Lucas's B.E.2c was in a mid-air collision with another aircraft of the same type flown by Second Lieutenant Geoffrey Jacques; both men were killed.[1]

There was debate – as there was for all of these academics-cum-pilots – as to the sensibility of allowing Lucas to train as a pilot. In the interests of the nation, should the lives of irreplaceable academics be risked in such a manner? Lucas's own view was that whatever the risk, men who are doing design and research work should experience the reality of the environment in which the output from that work will be applied. This sentiment was echoed by Melvill Jones, O'Gorman, and A.V. Hill in their posthumous tributes to Lucas.

The importance of Lucas's work in aeronautics is often overshadowed by his eminence as a physiologist, but archival material exists that sheds more light on his work at Farnborough (see the list of archives on p. 247 for details). A letter from O'Gorman to the ACA following Lucas's accident gives a feel for the respect he was

[1] There is a more detailed description of events preceding Lucas's death, and the crash itself, in McComas (2011, 63–74).

afforded by those working at the Royal Aircraft Factory (O'Gorman 1916a):

> On all the multiplicity of instruments designed by the Royal Aircraft Factory his advice was given unsparingly. His Gyro-Bomb-Sight had already far eclipsed every device of the kind & promised an approach to the accuracy of gun fire. He helped with aneroids, with speed indicators, with pressure tube measurements, with aeroplane levels & many other things, as well as on the aeroplane compass with which he has been chiefly associated. For months on end he rose at dawn to do the flying tests which his instruments were facilitating & eventually owing to the scarcity of pilots with scientific training in physics he asked to fly himself & was permitted to learn.

COMPASS ISSUES

In developing his superior navigational instrument, Lucas first compiled a report on the errors being recorded by existing aircraft compasses. Royal Aircraft Factory 'Report no. 251' provided a comprehensive account of the types of error peculiar to aviation compasses versus marine compasses. Vibration was the first error to be addressed, and the B.E.2 family of aircraft was chosen as the test platform. A standard 'pattern 200' compass was tested first, at various engine speeds, with the aircraft stationary, and it was found to be in error by as much as a worrying 42 degrees at 1600 revolutions per minute. Even worse, when two compasses were tested simultaneously in flight, one in the pilot's cockpit and one with the observer, the two differed from each other by between 40 and 50 degrees! The vibration experienced in one cockpit was clearly significantly different to that experienced in the other, something that was probably exacerbated by the proximity or otherwise of metal structures that might influence the magnetic field in the respective settings. A defined flight path was set up from a known point at Caesar's Camp to another point at the Royal Aircraft Factory.[2] It was established using this known track that the pilot's compass was no more than 10 degrees adrift; the observer's was the device giving the most erroneous information. The observer's seat was closer to the engine on the B.E.2, so the compass was deemed

[2]Caesar's Camp is an Iron Age hill fortification that is easily identifiable from the air and relatively close to Farnborough.

Figure 6.2 The Lucas Air Compass.

useless in the front cockpit but perfectly adequate if carried in the pilot's seat to the rear.

Next, the compass pivot was examined. The report reveals that 'Mr Darwin' had been involved. Horace Darwin had designed a circular vibration table, and results from his experiments clearly indicated that vibrational errors could be significantly reduced simply by inverting the pivot. Further tests revealed that combining this

Figure 6.3 Keith Lucas and his son David.

change with changes to the suspension resulted in errors being almost completely eliminated. As Lucas describes (Lucas 1914):

> These experiments suggested that the best suspension would be one in which the direction of the bowl was maintained by light springs, while some form of damping was used to prevent excessive oscillation of the bowl. In the form finally reached the bottom of the bowl was formed of a polished brass plate four and a half inches in diameter, which rested on a flat disc of felt half an inch thick. The bowl was supported from the sides of the case by flat leaf springs, three inches long, which pressed against flat faces on the inside of the case. These springs kept the bowl properly directed but allowed it to slide on the felt disc. The friction on the felt disc damped out excessive oscillations.

The importance of these experiments was not lost on Lucas or O'Gorman, who made sure the resulting design improvements were well documented in H-Department's 'Report no. 537' (O'Gorman 1916b). One of the first compasses incorporating the new design features is held as part of the Lucas family archive and can be seen in figure 6.2. Figure 6.3 is a precious photograph held in the same archive that shows Keith Lucas and his son David enjoying some time together not long before the accident that took Keith's life.

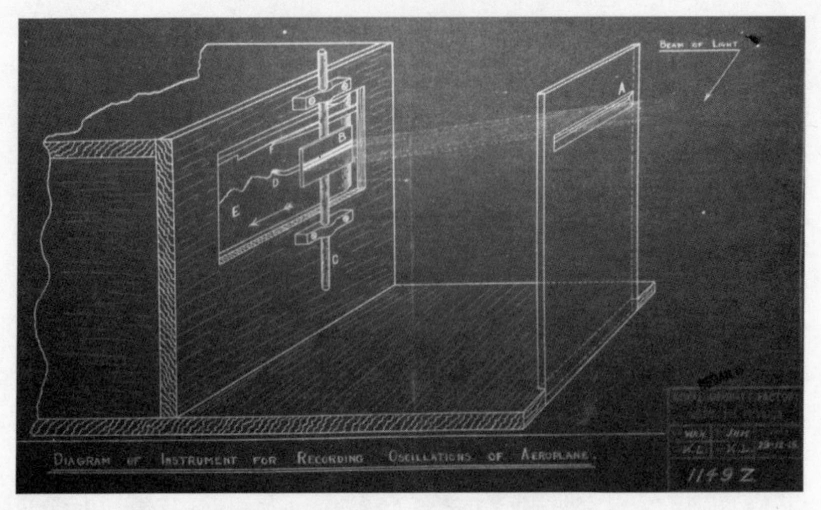

Figure 6.4 The photokymograph.

Rather fittingly, David Keith-Lucas[3] went on to study engineering at Gonville and Caius College, Cambridge, subsequently enjoying an influential career in aeronautical engineering, most notably with Short Brothers and at Cranfield Institute of Technology.

OSCILLATIONS AND THE PHOTOKYMOGRAPH

'Report no. 819a' from the Factory addressed tests conducted on aircraft oscillations. Rudimentary bomb-sights had been fitted to aircraft, but the many factors affecting the accuracy of weapon delivery were proving difficult to quantify and, hence, to correct. Oscillations of the delivery platform at the point of bomb release were certainly a factor, and understanding their magnitude, nature, and impact was therefore essential if progress was to be made towards accurate targeting.

The B.E.2c was again the aircraft chosen for conducting the experiments, with angular motion being recorded about all three aircraft axes in roll, pitch, and yaw. To record data, Lucas simplified a technique used by Busk in his stability investigations; his rather basic, but effective, apparatus is shown in figure 6.4, and he described it thus:

[3]Following Keith's death, the family decided to change their surname to Keith-Lucas as a mark of respect.

> A slit 0.6mm wide and about 10cm long is cut in a brass plate, and the plate is so fixed in the aeroplane that the slit is parallel to the axis of the motion to be recorded, for example, horizontal and athwartships when pitch is to be studied ... The direct light of the sun passes through the slit A and falls on the white plate B on which a black line is ruled parallel to the band of light. The plate B is moved by hand on the guide rod C so that as the aeroplane oscillates the black line is always kept in the middle of the band of light. The guide rod carries a style D which traces a curve on the metallic paper E moved by clockwork in the direction of the arrow at right angles to the motion of the guide rod.

The experiments using the photokymograph resulted in a much better understanding of how wind gusts or intentional control inputs disturbed an aircraft's longitudinal flight path. Lucas's grasp of mathematics is also apparent in his post-trial reports. The bulk of the mathematics appears in ACA 'Technical Report no. 118', and he summarizes his main findings in H-Department's 'Report no. 672' (Lucas 1915b):

> Any small longitudinal disturbance ... will cause an aeroplane to describe a path which is to a very near approximation a logarithmically damped sine curve. During this motion the inclination of the aeroplane to the horizontal, and its forward and angular velocities will vary harmonically with a period equal to that of the actual path, but with different phases ... In the first few seconds the resulting motion depends very largely on the way in which the aeroplane is disturbed, but after about a quarter of a period - roughly 5 seconds - the oscillations become simple damped harmonics.

A graph showing these damped pitch oscillations is contained in the Lucas archive (see figure 6.5): a source that also reveals similar investigations carried out to assess the nature of disturbances in roll (see figure 6.6). The character and depth of Lucas's analysis indicate something of the underpinning of his work as a physiologist. One expects a firm grasp of biochemistry in such a profession, but clearly Cambridge had also given Lucas a very good grounding in mathematics during his studies in natural sciences.

There is a letter in the archive that reveals something about Lucas's feelings and motivation for the job at hand. The intended recipient is not clear, nor are the particular circumstances in Germany that prompted the outburst, but it was penned by Lucas on 25 March 1915 and is a reply to someone who had asked him to

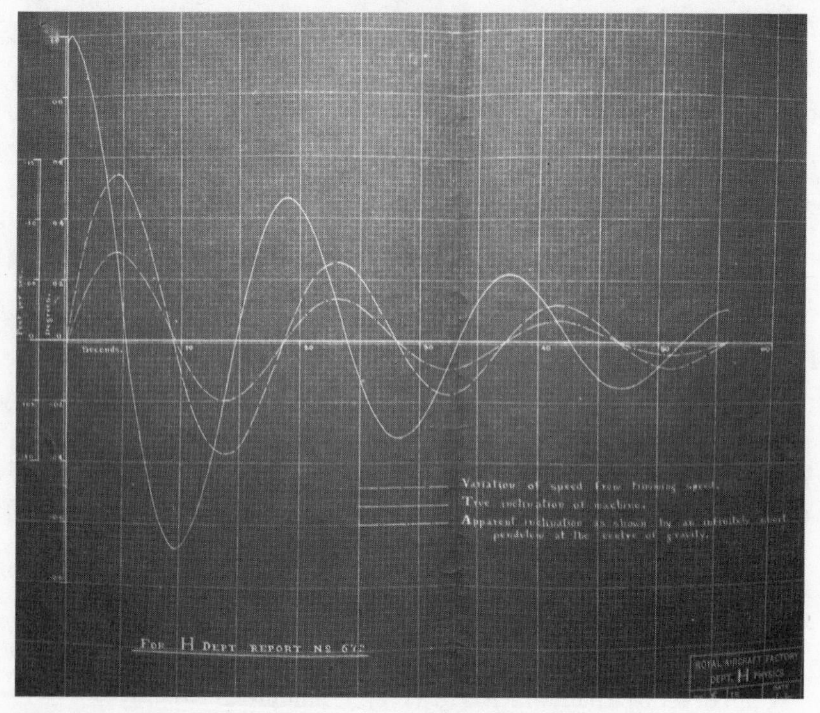

Figure 6.5 Aircraft speed and pitch oscillations.

review an academic paper. It is really quite a bitter and aggressive missive, expounding his views on the state of science and the corruption of scientific method in Germany. He appears to have particular incidents and individuals in mind, but he uses generic terminology to avoid explicit accusation (Lucas 1915a):

> I have most interesting work at the Royal Aircraft Factory designing and testing new devices. I get a lot of flying, usually two or three times a week, and I believe the work I have done has been of some use … It pleases me to hear you say that this is some of your fight too. I believe it is. If I didn't believe it was the fight of every one who cares for freedom I shouldn't be tootling about the sky in aeroplanes. It is my own conviction that in science as much as in politics this is a fight for freedom. The country in which V. smashed the apparatus put up by one of his underlings to test a question which V. had not suggested, or where F. could not write the results of his researches on 'all or none' until he left the University of X, where Y was professor and held opposing views; the country where 'Es ist leicht zu sehen' [It is easy to see] and 'Man muss annehmen' [You must accept] take the place of observation,

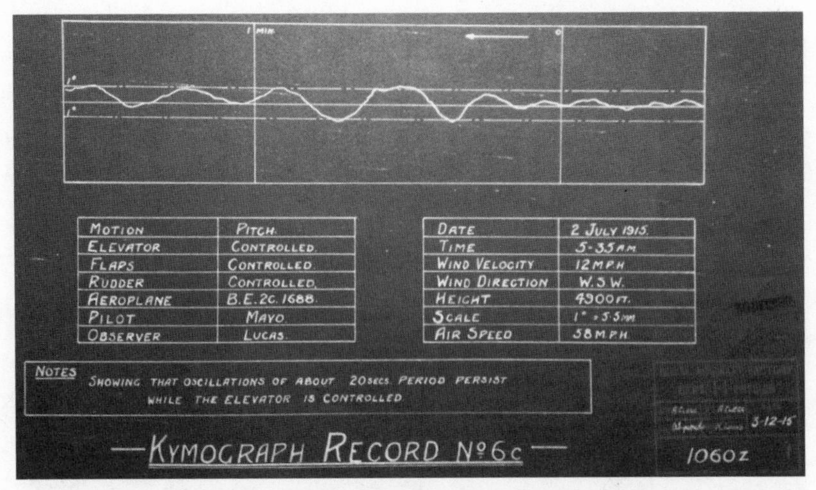

Figure 6.6 Aircraft roll oscillations recorded using the photokymograph.

> and the professors set up a hierarchy of science ... This is my view of the case. I fear German *scientismus* as much as German *miltarismus*, and I believe the origin of both the same. Send me your paper. It is a relief to hear from a place where truth still ranks before gain.

Lucas's conviction to the cause is clear, and it is a stance that would sadly lead to his untimely death.

THE LOGBOOK

As an epilogue to the scientific exploits and achievements of Keith Lucas during his time at the Royal Aircraft Factory, it is perhaps worth contemplating the contents of his flying logbook. They reflect the very short duration of his flying career as a pilot: three double-page entries tell the story of his brief quest to gain his wings. Figure 6.7 shows the first double page. He received 2 hours and 56 minutes of familiarization in the first week and, after just 1 hour and 40 minutes of flying during the following week, his instructor, Lieutenant Jarvis, decided that Lucas was ready for his first solo: a memorable moment in any pilot's career.

Lucas completed one circuit of the aerodrome and landed successfully. By the end of the week he had accumulated more than 3 hours of solo flying, and his flying experience totals 7 hours and 48 minutes. In week three, Lucas was deemed competent enough to

Date and Hour	Wind Direction and Velocity	Machine Type and No.	Passenger	Time	Height	Course	Remarks
		M.F					
9.9.16/10.40	ENE.15	A.2438		20′	500′	aerodrome	Began instr. under Capt. Yates.
6. pm	NE 15	A.952		15′	1300′	aerodrome	Instr. under Sgt. [illegible]
10.9.16/6.40AM	NE 10	A 952		21′	500′	aerodrome	ditto. 2 landings
11.9.16/10.15AM	W 5	A 952		21′	500′	aerodrome	took [illegible] 1st time. 3 landings [illegible]
11.30AM	W 10	A 2178		20′	500′	aerodrome	with [illegible] for air experience
7.10PM	NW10	A 2178		20′	700′	aerodrome	practiced landings with Lt. [illegible]
12.9.16/8.10AM	NW 10	A 953		10′	700′	aerodrome	practiced landings with Sgt. [illegible]
6.50pm	NW10	A 2438		24′	500′	aerodrome	practiced landings
13/9/16. 8.0AM	NW 20	A2438		25′	600′	C.F.S.	with Sgt. Chestwood.
			2 hrs	56 min			

Total time in air during week 2 hrs 56 min
Total time solo for week nil
Grand total solo nil
Grand total time in air 2 hrs 56 min

Figure 6.7 Flying logbook of Keith Lucas: first week entries.

fly the B.E.2 with an instructor. The final entry in his logbook was recorded on 26 September 1916, with Lucas's experience in the air totalling 11 hours and 18 minutes: he was now ready to go solo in the B.E.2. We do not know exactly how many hours he had accumulated by 5 October, but it would have been fewer than 20 at the point at which the mid-air collision took his life.

THE PHYSICIST

Frederick Lindemann was the scientist who was most ardent in his resolve to convince the hierarchy that people like him should be allowed to fly. Like Taylor and Lucas, Lindemann was frustrated by the military pilots, who demonstrated little empathy with the scientific or mathematical principles under investigation; nor could they speak the jargon of the scientist. In the kerfuffle over the Factory's role in mass production of airframes and engines and O'Gorman's movement sideways from Farnborough into a post at the War Office, Lindemann seized the opportunity to convince the new supremo, Henry Fowler, that pilot training was in order, and it was agreed that half a dozen or so academics should be taught to fly.

It was the summer of 1916 when Lindemann, Lucas, Thomson, and Farren arrived at CFS on Salisbury Plain, which was then under the command of Henry Thomas Tizard, a mathematician and chemist, and one of the few people in this story to be educated at Oxford. Tizard joined the RFC via the Royal Garrison Artillery and was a pilot very much in the mould of Busk: someone who enjoyed conducting his own flight experiments, with a keen understanding of the mathematics and physics supporting his observations.

Tizard was Hopkinson's deputy at the time of the latter's untimely death and became a very influential character in British aeronautics towards the end of the war. He joined the Royal Air Force when it was formed in 1918 but soon left to return to his passion of chemistry, renewing his links with Oxford. Lindemann's arrival at CFS reunited Tizard with a close friend, the two having met in 1908 in Berlin, where they formed a strong bond. Knowing this, a reasonable assumption might be that Tizard had some hand in the freeing of Lindemann (and others) from Farnborough's tight clutches.

Fortunately, Lindemann's flying logbooks survive among the contents of his personal archive, and they provide us with invaluable insight into the nature and extent of his flying experiences.[4] Two logbooks exist: one for his time under training, the other documenting his flying at Farnborough. We learn that he began flight training with No. 8 Reserve Aeroplane Squadron on 9 September 1916, with his first solo flight following in short order on 16 September; he was at the controls of a B.E.2c less than a week later. The end of October saw him graduate and return to the Royal Aircraft Factory with his academic friends, minus Lucas, who had by then perished in the mid-air collision.

Lindemann's Farnborough logbook tells us that his first test flights at the Royal Aircraft Factory began on 21 November 1916. Paired up with Hugh Renwick, the two would conduct 32 flights together testing a new aeronautical compass; one imagines that had Lucas survived, this test programme would have been delegated to him, as he was the compass designer. In Renwick's absence, H. Grinstead, another of the Chudleigh lot, took the observer's seat during the trials.

[4]Lindemann's extensive personal papers are held at Nuffield College, Oxford.

Figure 6.8 Frederick Lindemann (pillion) sporting his distinctive bowler hat.

Grinstead proved to be one of the most elusive of the Chudleigh group. There is little information available to determine his background or establish his exact role at the Factory; he is certainly a missing piece of the puzzle in terms of this story, and it is frustrating not to be in a position to give him appropriate and commensurate recognition. What is known is that both Grinstead (2 December 1916) and Renwick (5 January 1917) experienced and survived aircraft in-flight engine failures with Lindemann at the controls: an indication both that aero engines were prone to failure at that time and that Lindemann clearly knew how to 'dead-stick' an aircraft onto the ground safely. If a powered aircraft loses all propulsion, the pilot is said to be flying with a 'dead stick', and the aircraft effectively becomes a glider. Sudden engine failure on a single-engined aircraft at slow speed requires swift initial action to avoid a stall: one must rapidly push the nose down to maintain flying speed while there is still elevator authority. Lindemann's logbook entries were all countersigned by Frank Goodden, the chief test pilot at the Royal Aircraft Factory, who oversaw test flying during this period.

ACCELEROMETER TEST: 'REPORT NO. 621'

Once the compass tests were complete, Lindemann was assigned duties that involved investigating the performance of a number of other aircraft instruments and operational aids. Reports from these studies are very revealing in a mathematical sense. Accelerometer tests, for example, were summarized in H-Department's 'Report no. 621' (Lindemann 1917b), the first page of which can be seen in figure 6.9. The critical point Lindemann makes in this particular piece of work is that the bumps an aircraft experiences during a flight are short and sharp, so any device being employed to measure them that uses the movement of a mass in its action will have to ensure that that mass is heavily damped. In his analysis, he uses the standard second-order differential equation that describes simple harmonic motion to derive an expression for the displacement of the indicator, and he explains why the mass used needs to be relatively small.

It is noted that Cambridge mathematician David Pinsent, who will be discussed in chapter 8, flew as an observer on one such test flight on 29 April 1917, so he may well have been involved in assisting Lindemann with the mathematical analysis. Figure 6.10 shows an actual test output from the device; the graduation down the left-hand side indicates the number of 'g'. The output shown is a record of the forces being experienced during a loop manoeuvre. It is evident that Lindemann actually pulled about 2.6g entering the manoeuvre. He was almost weightless over the top of the loop and then pulled 2g on the recovery. In terms of aircraft performance, Lindemann would have entered the loop at about 95 miles per hour and been at very low speed indeed over the top.

Information regarding the physical nature of the accelerometer is given by H.B. Howard. A peripheral player in our story, Howard was another Cambridge graduate who came down in 1916 straight into a job in the technical branch of the Air Department at the Admiralty under Harris Booth. Howard described Booth as an 'erratic

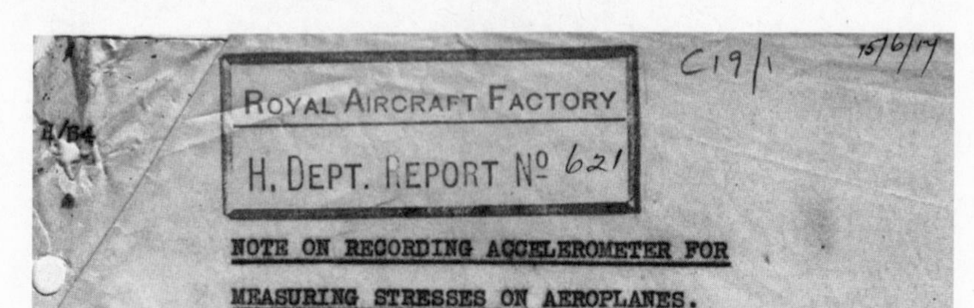
Royal Aircraft Factory
H. Dept. Report No 621

C19/1 15/6/17

NOTE ON RECORDING ACCELEROMETER FOR MEASURING STRESSES ON AEROPLANES.

Stresses due to accelerations may be measured by means of any arrangement capable of recording the displacement of a mass from a position of stable equilibrium. For use on aeroplanes where very short sudden accelerations occur, e.g. in bumps, landing, taxiing, etc., it is essential that the period of the moving parts should be short and that the damping should be great, if possible so large that the movement is practically aperiodic. In any instrument in which the mass is held in its position of equilibrium by an elastic force the displacement for a given acceleration depends only upon the period. For if, neglecting damping,

$m \frac{d^2x}{dt^2} = -Rx$ defines the mass m and the restoring force R, the displacement $x = -\frac{m}{R}\frac{d^2x}{dt^2} = \frac{\tau^2}{4\pi^2}\frac{d^2x}{dt^2}$ where τ is the period. Thus if τ is to be short x must be small and some magnifying arrangements must be used. The logarithmic decrement is proportional to the frictional force and inversely proportional to the mass of the moving parts. The best way to make it large while keeping the instrument small and light is to use moving parts of very small mass.

Figure 6.9 H-Department's 'Report no. 621' by F.A. Lindemann from June 1917: accelerometer test notes.

genius'. Booth's first question at Howard's job interview was rather indicative: 'Can you work when you're fed up, because you will get fed up?' (Howard 1966, 65). One wonders if Booth was implying the job was tedious or simply frustrating. Recalling his time working in

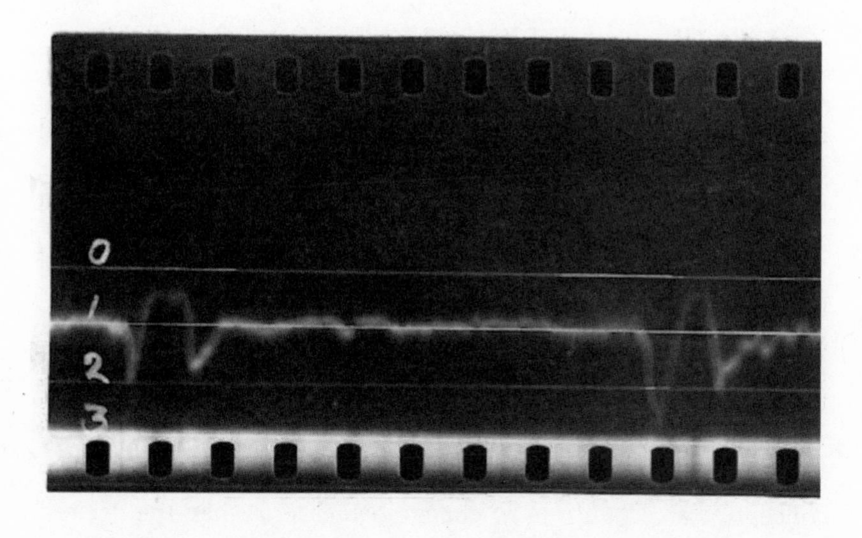

Figure 6.10 H-Department's 'Report no. 621' by F.A. Lindemann from June, 1917: accelerometer stall and loop results.

conjunction with the Royal Aircraft Factory, Howard describes the accelerometer that Lindemann used (Howard 1966, 59):

> The detecting unit was a very fine glass fibre bent into a semicircle of about 1 inch in diameter, with the two ends clamped. The tip of the fibre was illuminated and its motion recorded on film. The glass fibre thus formed both the mass and the spring with a period of about one-twentieth of a second, well away from that of the loads it was to record. Air damping was increased by enclosing the unit in a small chamber.

Howard also talks about calculations that had been done in 1913 at Farnborough by Thomas Wigston Kinglake Clarke, who had used standard equations of motion to analyse the theoretical forces that might be experienced by a B.E.2c when pulling out of a vertical dive. The figures he derived were 9.5g for a drop of 1,000 feet, and 11.5g if pulling out of a dive that had reached terminal velocity. Given that 6g was at the top end of design parameters, it is hardly surprising there were so many structural failures in the early days of aviation!

RATE-OF-CLIMB INDICATOR: 'NOTE ON R&M NO. 310'

Another set of tests investigated the performance of a rate-of-climb indicator, as described in 'Note on R&M no. 310' (Lindemann 1917a) and shown in figures 6.11 and 6.12. The report illustrates the typical mathematics being undertaken by Lindemann in relation to aircraft instruments.

Most aircraft of the time were fitted with a rudimentary artificial horizon, a compass, a rate-of-turn indicator, a rev counter for the engine, and an altimeter. In many respects, both the accelerometer and the rate-of-climb indicator were instruments that were more for the experimentalists than for the front-line aircraft pilots.

Again, the mathematical analysis of the operation of the rate-of-climb indicator hinged upon differential calculus and the familiar concept of a certain rate of change being zero when the desired constant motion was achieved. An interesting observation from the first page of 'Note on R&M no. 310' is the clear awareness of the need to keep weight to an absolute minimum when designing aircraft instruments: with the limitations on aero-engine power output, power-to-weight always had to be at the forefront of a designer's mind.

BOMB-SIGHT TRIALS

Several sorties were used to evaluate a bomb-sight designed to be employed in a diving trajectory. The question raised was this: could more accurate bombing be achieved in a dive rather than in straight and level flight? The B.E.2e trials indicated that diving did not improve accuracy, a conclusion noted in H-Department's 'Report no. 626'. This is quite a surprising conclusion, particularly when one considers the success of the Luftwaffe's Junkers Ju 87 'Stuka' dive-bomber during World War II. Lindemann reported issues with oil from the engine covering his goggles during a dive, so perhaps it was the lack of vision that scuppered the accuracy rather than the technique.

One gets a real sense of life as a flying mathematician from the entries in Lindemann's logbook: the diversity of the trials, the variation in crew, and the constant requirement to write up reports on the ground and make sense of the mathematics and physical principles being investigated. Pinsent and Renwick were Lindemann's most frequent companions in the observer's seat. The final mention

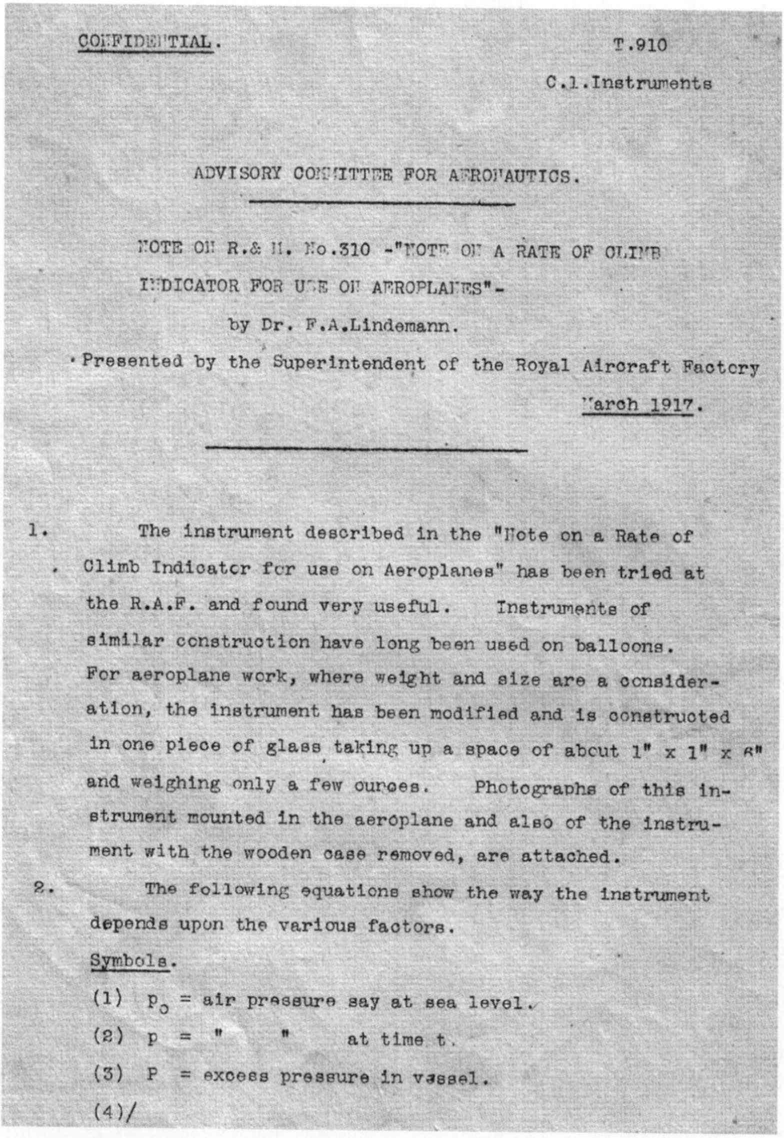

CONFIDENTIAL. T.910

C.1.Instruments

ADVISORY COMMITTEE FOR AERONAUTICS.

NOTE ON R.& M. No.310 -"NOTE ON A RATE OF CLIMB INDICATOR FOR USE ON AEROPLANES"-

by Dr. F.A.Lindemann.

Presented by the Superintendent of the Royal Aircraft Factory

March 1917.

1. The instrument described in the "Note on a Rate of Climb Indicator for use on Aeroplanes" has been tried at the R.A.F. and found very useful. Instruments of similar construction have long been used on balloons. For aeroplane work, where weight and size are a consideration, the instrument has been modified and is constructed in one piece of glass taking up a space of about 1" x 1" x 8" and weighing only a few ounces. Photographs of this instrument mounted in the aeroplane and also of the instrument with the wooden case removed, are attached.

2. The following equations show the way the instrument depends upon the various factors.

Symbols.

(1) p_0 = air pressure say at sea level.

(2) p = " " at time t.

(3) P = excess pressure in vessel.

(4)/

Figure 6.11 'Note on R&M no. 310' by F.A. Lindemann from March 1917: rate-of-climb indicator outline.

of Pinsent appears in an entry for 27 April 1918, when he took to the air to conduct a range-finding trial using an R.T.1 aircraft, while Renwick's last recorded flight in this role came on 15 August 1918: a rudder gyro control test in an F.E.2b aircraft.

If equation (1) is written

$$a \frac{dP}{dt} + b P \frac{dP}{dt} + cP + d = 0$$

then

$$t = \frac{bd-ac}{c^2} \log (d + cP) - \frac{b}{c} P + \text{const.}$$

or

$$t_2 - t_1 = \frac{bd - ac}{c^2} \log \frac{d + cP_1}{d + cP_2} + \frac{b}{c} (P_1 - P_2) \qquad \text{(II)}$$

This defines the time taken to settle down to a new value P_2 when the rate of climb changes. It is obviously of the same order as the time taken for P to sink to 0 at constant p.

The equilibrium value of P at a constant rate of climb is found by putting $\frac{\delta P}{\delta t} = 0$. This gives

$$P = \frac{-v_0 \frac{dp}{dt}}{+ \frac{\pi r^4 p}{8 l \eta} \pm \frac{q}{\rho g} \frac{dp}{dt}} = \frac{+ \rho g \; v_0 \tau \frac{dp}{dt}}{+ \rho g \left\{ (v_0 + q\, h_0) \log 2 + 2 q H_0 \right\} + \tau q \frac{dp}{dt}}$$

or

$$H = \frac{v_0 \tau \frac{dp}{dt}}{\rho g \left\{ (v_0 + q\, h_0) \log 2 + 2q\, H_0 \right\} - \tau q \frac{dp}{dt}} \qquad \text{(III)}$$

Figure 6.12 'Note on R&M no. 310' by F.A. Lindemann from March 1917: the mathematics of the rate-of-climb indicator.

Amid the wealth of information contained within it, the logbook not only tells us the story of a pilot who lived through the war, but also provides insightful snippets about his colleagues. A further revelation is that pilots were sometimes flying with each other rather than always being paired up with non-pilot observers; flights with staff from the Stressing Section at the Admiralty are also recorded. Thomson and Taylor from the Factory, and structural engineer Laurence Pritchard from the Admiralty, are all listed as flying with Lindemann on gyro test flights.

Pinsent's respect for Lindemann verged on awe. In an undated letter to his mother soon after moving to Chudleigh, he excitedly lauds Lindemann for his ability to perform mathematical

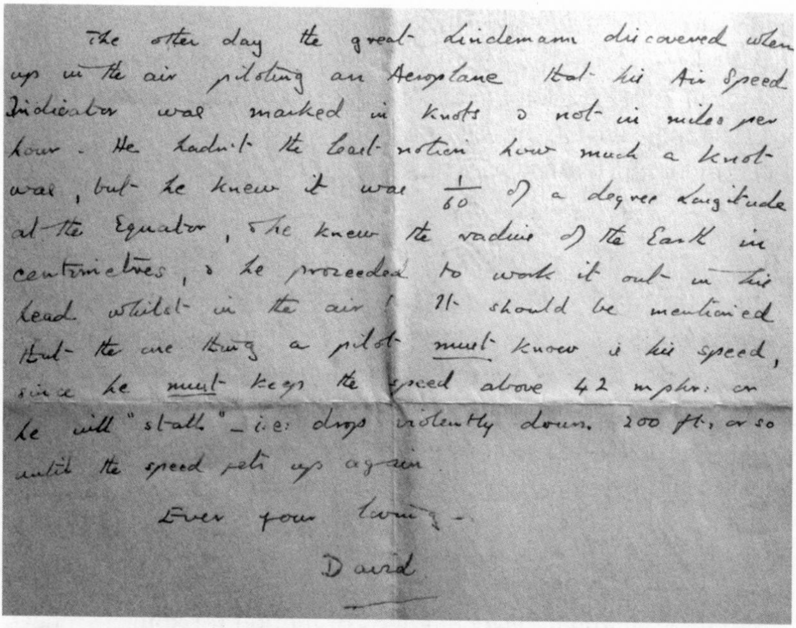

The other day the great Lindemann discovered when up in the air piloting an Aeroplane that his Air Speed Indicator was marked in knots & not in miles per hour. He hadn't the least notion how much a knot was, but he knew it was $\frac{1}{60}$ of a degree Longitude at the Equator, & he knew the radius of the Earth in centimetres, & he proceeded to work it out in his head whilst in the air! It should be mentioned that the one thing a pilot <u>must</u> know is his speed, since he <u>must</u> keep the speed above 42 mph: or he will "stall" – i.e: drop violently down 200 ft, or so until the speed gets up again.

Ever your loving –

David

Figure 6.13 Pinsent writes of Lindemann's exploits.

mental gymnastics while in the air (the actual note is reproduced in figure 6.13):

> The other day the great Lindemann discovered when up in the air piloting an Aeroplane that his Air Speed Indicator was marked in knots and not in miles per hour. He hadn't the least notion how much a knot was, but he knew it was $\frac{1}{60}$ of a degree longitude at the Equator, and he knew the radius of the Earth in centimetres, and he proceeded to work it out in his head whilst in the air! It should be mentioned that the one thing a pilot <u>must</u> know is his speed, since he <u>must</u> keep the speed above 42 mph or he will 'stall' – i.e. drop violently down 200 ft, or so until the speed picks up again.
>
> Ever your loving,
>
> David

SPINNING

Lindemann had it in mind to sort out the problems surrounding the spinning characteristics of aircraft at the earliest opportunity. It was not an aspiration to be taken lightly. Lindemann was still relatively fresh out of pilot training at this point, with very little experience compared with the likes of Goodden. He had watched others plummet to their deaths in spins and he had written down in mathematical terms what he thought was happening. Furthermore, by studying the implications of his analysis, he was able to specify and justify the exact technique he believed was needed to rectify the problem.

A spin is a type of stall characterized by the aircraft dropping downwards while autorotating about its vertical axis. A normal spin is initiated when one wing stalls while the other is still producing lift. This can easily happen if, when approaching the stall angle, some yaw is present. An aircraft in such a spin will have a high angle of attack, it will have an airspeed below stall speed as sensed by at least one of the wings, and it will be in a shallow descent while automatically rotating about its vertical axis. This spin should not be confused with a spiral dive: a situation in which the wings are not stalled and the aircraft responds normally to control inputs.

A raft of mathematical equations of motion for spins had been laid out by George Thomson as early as 1915 (Thomson 1915). Thomson analysed an aircraft in a sideslip during a banked turn: a classic configuration to encourage an incipient spin and one often encountered by pilots. When attempting to land an aircraft, it is common practice to fly 'downwind' parallel to, and laterally displaced from, the runway. After passing abeam the threshold of the runway, a banked turn is commenced to bring the aircraft around through 180 degrees to line up on the centreline, into the wind. If the pilot misjudges the turn by not applying enough bank, or by failing to allow correctly for any crosswind, it is easy to pass through the centreline and risk not hitting the runway at the desired point. In such circumstances, pilots will often attempt to compensate for such an error in the latter stages of the turn by applying greater bank and adding rudder to skew the nose around. This is now the perfect set of conditions for potentially leading to disaster: the aircraft is low, slow, over-banked, and yawing – right on the verge of flicking into a spin. With the ground in such close proximity, there is no hope of recovery if this happens. This 'mishandling in the finals turn', as it is now known, is how many of the early spinning fatalities occurred. Another potentially fatal scenario was when pilots attempted an immediate turn back to the runway following an engine failure after take off.

Returning to Lindemann, while he and the other mathematicians had been away learning to fly, the new F.E.8, a single-seat scout aircraft, had been causing all manner of distress as pilot after pilot met his death, each one unable to execute an effective spin recovery. The B.E.8, a new two-seat recce aircraft, was encountering similar issues. In his fond reminiscences of his time at Chudleigh, Cambridge graduate Ronald McKinnon Wood identifies pilots Goodden and Harry Hawker, the latter employed by Tom Sopwith at the time, as having done their utmost to fathom the spin recovery. Despite being able to execute the recovery successfully, though, they were still unsure about what precisely was going on with the aircraft (McKinnon Wood 1960, 1533).

As an aside here, Harry George Hawker was an Australian pioneer of aviation and a test pilot. His name is perhaps best known for the World War II fighter aircraft that also bore it: the Hawker Hurricane. The aircraft was produced by the Hawker Aircraft Company, although Hawker himself died many years previously in an air crash.

Thomas Octave Murdoch Sopwith came into aviation in 1906 via Short Brothers, who were making balloons in Battersea at the time. It would be watching the American John Moisant performing some exhibition flying in Folkestone in 1910 that would inspire Sopwith's interest in powered flight. Within two months of watching Moisant's antics, Sopwith had taught himself to fly and held a flying licence: the 31st issued in Britain. In 1912, he set up the Sopwith Aviation Company with a number of other aviation enthusiasts. The company would go on to produce the famous and influential Sopwith Camel single-seat fighter aircraft in vast numbers for the RFC.

Ronald McKinnon Wood arrived at Chudleigh during the latter part of the war, and he eventually replaced William Farren as the head of the aerodynamics department at the Royal Aircraft Factory when the latter moved on. The former played a significant role in reestablishing links with those involved in Germany's aerodynamics research following World War I, particularly Ludwig Prandtl, Albert Betz, and Max Monk.

The first recorded spin recovery in Britain had been made by Lieutenant Wilfred Parke in August 1912 while flying an 'Avro Type G' cabin biplane. Rather fine margins were involved in Parke's escape, with luck rather than judgement resulting in a spin recovery barely 50 feet above the ground. This feat became somewhat legendary, and its essence is captured by Berriman in an article entitled 'Parke's dive' in the August 1912 issue of *Flight* (Berriman 1912b). Berriman had been watching the aircraft being put through its paces earlier but had adjourned for breakfast when the incident occurred; he did, however, witness the shaken pilot marching off with de Havilland and Frank Short for what must have been a rather interesting post-flight debrief, details from which allowed Berriman to reconstruct and record the incident for posterity. Short was a Royal Aircraft Factory engineer who designed the air speed indicator used by most early British military aircraft. He called it a 'velometer'.

Remarkably, it would be Lindemann who carried out the first practical scientific analysis of the spin. Unwilling to let a military pilot follow his instructions, he elected to put his own life on the line and carried out a sequence of deliberately induced spins. As physicist R.V. Jones later put it, 'there is little doubt that Lindemann was the first man, at least in Britain, to put an aircraft into a spin and then recover it knowing what he was doing' (Jones 1987, 192).

Reginald Victor Jones studied natural sciences at Wadham College, Oxford. He graduated in 1932 and was immediately championed by Lindemann and awarded a research studentship that gave him expertise in the science of infrared detectors. Jones, encouraged by Lindemann, moved into aeronautics, and he went on to develop the technology required for airborne infrared detectors, as used by British night fighters during World War II. Supervised by Henry Tizard, Jones's work was conducted at the Clarendon Laboratory in Oxford.[5]

The spin tests began on 11 June 1917 using a B.E.2e, with Lindemann throwing the aircraft into a spin from 7,000 feet and recovering before encountering anything solid, making mental notes of nine different parameters during each test. Lindemann memorized the air speed, the yaw, the angle of attack of each wing (measured using ribbons attached to the wing struts), the height before and after, the number of turns executed, and the elapsed time. Following each plummet he would write down everything he recalled, thus building up a data set that allowed him to back up his theoretical assertions with actual numbers. Sometimes he would induce the spin above a ground-based camera obscura to enable feedback on parameters such as radius of turn (Brinkworth 2014, 113).

[5] Jones's extensive collection of papers and correspondence is held in the archives of Churchill College, Cambridge.

Initial research suggested that Lindemann flew all these spin tests solo, but his flying logbook tells a different story. After the initial spinning test flight, which certainly was solo, subsequent sorties of this nature, mostly in the 'c' variant of the B.E.2, had an observer present. Those listed in the logbook are Stevens, Pinsent, Grinstead, and Renwick, the latter noting down the data from the final tests on 27 July 1917, initiated at 6,000 feet in a B.E.2c. H.L. Stevens had joined the team at the Royal Aircraft Factory directly from Cambridge in 1914 and was assigned a role working under Busk. Perhaps his most significant work was designing the aerofoil wings that would lift many of the Factory's aircraft into the air.

The presence of another person in the aircraft in no way diminishes the status of Lindemann's achievement, but it does say much for the bravery of the observers. In many ways, Lindemann's actions were just as special as Busk's a few years earlier: he was an academic who was able to look at a physical event, translate it into mathematics, and then use this mathematics to suggest a physical solution to a problem. One must also stress the remarkable courage, composure, and mental agility that would have been required to execute trials of this nature. Lindemann left the Royal Aircraft Factory at the end of the war to take up a professorship in experimental philosophy at Oxford, and during the interwar years he became a close friend of Winston Churchill. Upon becoming prime minister in 1940, Churchill appointed Lindemann as the British government's senior scientific advisor.

The roles and contributions of the observers on these flights should also be acknowledged. Being a passenger in an aircraft in a spin is somewhat unnerving at the best of times, so remaining calm enough to accurately monitor the spin characteristics is impressive, particularly in an open cockpit. Lindemann's achievement meant that a standard spin recovery procedure could be taught to all pilots: something that must have saved countless lives. It certainly saved me trauma on a number of occasions when my aerobatic sequence did not quite go to plan! Despite this crucial knowledge, however, and even towards the end of the war in May 1918, for example, 41 of the 106 recorded flying accidents involved spinning. It was clearly still a huge issue, therefore, and it was not helped by a frustrating delay in the formal publication and dissemination of Lindemann's findings and recommendations (Lindemann, Glauert, and Harris 1918).

Lindemann – a peripatetic academic aviator wandering between various projects – certainly worked on many things during his time at Farnborough.[6] He even found time to undertake mathematics unrelated to the flying task at Farnborough, with his suggested geometrical method of rectifying the arc of a circle illustrating the breadth of his mathematical interests, which complemented his passion for physics and chemistry. Piloting an aircraft on a mission to test the strength of an aircraft wing versus a balloon mooring cable was perhaps pushing things a little too far, however. Thankfully, Chief Test Pilot Roderic Hill was there to step in and save the day. Both Hill and this hair-raising incident will be discussed in more detail in chapter 8. Lindemann's final logbook entry – a test on a Sperry rudder control – was made after the war on 3 January 1919.[7]

[6] A full, informative, and entertaining account of Lindemann's time at the Royal Aircraft Factory can be found in chapter 3 of Fort (2003).

[7] The Sperry brothers, Elmer and Lawrence, were American aviation pioneers who specialized in automatic aircraft control systems. They developed a number of systems built around gyroscopes, and as early as 1916 they had created a rudimentary autopilot.

7

Industrialization and the Admiralty

THE INDUSTRIALIZATION OF AIR WARFARE

Industry identified that a lucrative market was opening up in aeronautics, so many aeronautically enlightened entrepreneurs moved to take advantage – individuals who would later become household names: Shorts, Handley Page, Roe, and Sopwith, to name just a few. A typical example of this kind of businessman was Richard Fairey. Perhaps it was the proud family heritage in the carriage-building trade that lay at the heart of his inspiration, but one suspects it was also the guiding hand of Silvanus P. Thompson at Finsbury Technical College that cajoled the young Fairey into the world of engineering. Thompson was the principal and a professor of physics at Finsbury Technical College for 30 years (1878–1908) and was a great proponent of technical education as a means of converting science into practical engineering.

Fairey's link to aeronautics came in 1910, when the 22-year-old Londoner inadvertently infringed a patent registered to Dunne while competing in a model aeroplane competition at Crystal Palace. This provoked a meeting between the two and, subsequently, Fairey's appointment in 1913 as chief engineer at Short Brothers, which was at the vanguard of the trade. Two years later, Commodore Murray Sueter, one of the military representatives on the original ACA, would convince Fairey to set up his own firm.[1] The rate of attrition suffered by the RFC as its German counterpart, the Fliegertruppen des deutschen Kaiserreiches, gained the upper hand in the skies over France during the earlier part of the war must have been a significant motivating factor in the British

[1] For a more comprehensive discussion of Fairey's life, see Adrian Smith's ongoing research project at the University of Southampton: 'The life of Sir Richard Fairey (1887–1956), aircraft designer and industrialist'.

military's proactivity in encouraging rapid expansion of industrial capacity in the private sector.

The Royal Aircraft Factory was owned by the taxpayer, and the initial intent was for it to be the exclusive centre for the design of military aircraft; only once the designs and prototypes were in place could subcontracting to the privateers be considered. But that is not how things materialized.

From the time the RFC was born in 1912 with its army strand and its navy strand, the former acquiesced but the latter elected to adopt a unilateral policy that allowed it to source the best aircraft available, whatever their origins. This meant that when war began, the RFC aircraft in France almost exclusively comprised Farnborough-designed aircraft, while the newly formed Royal Naval Air Service brimmed with the fruits of the privateers' labours.

In the autumn of 1914, the Admiralty ordered the first air raids over Germany, with its pilots flying a variety of aircraft designed and manufactured by the private sector. The aircraft used to target the German Zeppelin sheds came from Shorts, Avro, and Sopwith. A case could be made that these forays marked the start of the industrialization of air warfare in Britain. Up to this point, reconnaissance had arguably been the primary role of aircraft, but things were about to change, and substantially: the military wanted aircraft that could not only scout but also drop ordnance from the air and fire bullets when required. It was the latter issue that brought things to a head regarding the roles of the Royal Aircraft Factory and industry in providing the aircraft to do the job.

While those at Farnborough strove desperately to produce a capable combat aircraft with a pusher propeller, so that a machine gun could be fired forward, unobstructed, the Germans stole a march by employing an 'interrupter' mechanism that could synchronize bullet fire between the propeller blades of their tractor aircraft. The British efforts resulted in the F.E.2b aircraft from the Royal Aircraft Factory as well as a contribution from Vickers: the 'Gun-bus'. 'Pusher' aircraft had the propeller(s) behind the cockpit and therefore pushed the aircraft through the air, whereas 'tractor' aircraft had the propeller(s) in front of the cockpit, so the aircraft was pulled through the air.

By the end of 1915, the devastating losses on the British side caused public outcry. This was the tipping point. It was no longer practical or acceptable to restrict RFC hardware to that which the Royal Aircraft Factory could design and manufacture: the pilots on the front line needed the best equipment available, and much of that lay in the manufacturing sheds of the industrialists.

The experience and talent from the Royal Aircraft Factory had begun migrating into industry anyway by this point, with apprenticeships served at Farnborough forming the core experience in some of the major industrial design offices. Typical of this osmosis was Geoffrey de Havilland. The man who had been the initial inspiration at the Factory had moved on to form the Aircraft Manufacturing Company, and it was from here that the D.H.2 emerged: the British aircraft that would compete on an equal footing with the German fighters.

The scale of industrial aeronautical development at this time was remarkable. At the end of the first decade of the century, a few pioneering individuals were cobbling together aircraft in the corners of garages and workshops, but by 1915 more than 30 companies were engaged in aeronautical work. By the end of the war, Britain's aeronautical industry employed nearly 350,000 people.[2] In terms of aircraft production, numbers had risen from literally a handful in 1914 to over 26,000 during the last year of the war; engine production over the same year was upwards of 30,000 units (see Bowyer 1966, 282).

This is a story repeated throughout history, where conflict drives innovation and expansion. A hundred years earlier, stimulated by the navy's requirements for ships to fight in the French Wars, the first mass-production lines in the world were set up to make the blocks used in the rigging of the sails of British ships. Demand for cannons drove expansion in iron-working, and new tools such as the industrial lathe, invented by Henry Maudslay, appeared in 1797. Aeronautics and World War I brought similar advancements. This time aero engines were filling the production lines, the fabrics were for wings not sails, and the drawing offices were brimming with plans for airframes instead of ship hulls.

The armistice was a blessing for the world, but not for the aeronautical industrialists. The hiatus between military demands in wartime and civilian demands for transportation, which would increase exponentially a few years later, forced much industrial capacity to be mothballed. Regardless, Britain had transitioned

[2]A concise summary of the industrial expansion in British aeronautics can be found in Fearon (1969).

from the parochial nature of early aviation into an era of industrialization in aeronautics.

While many of the Chudleigh lot moved back into previous endeavours or on into new ones as the pressure for aeronautical advancement eased, one of their number elected to make aviation a career. William Farren was a true aviator. His passion was being in and around aircraft, and he would spend the entirety of his working life helping to advance British aeronautical know-how.

THE AVIATOR

William Scott Farren won a scholarship in 1901 to the Perse School, Cambridge, where he was taught mathematics by V.M. Turnbull, 14th Wrangler in 1891. Ten years on, and with a passion for mathematics and physics, Farren earned another scholarship, this time for Cambridge University, where he obtained first class honours in the Mechanical Sciences Tripos in 1914. Two contemporaries of his at that time were George Thomson and Robert Mayo: Thomson is known to have had a great deal of respect for Farren's instinctive approach to engineering, once saying of him, 'He looked at a machine as an animal, not as an example of various scientific laws' (Thomson and Hall 1971, 216).

Of all the flying mathematicians who worked at the Royal Aircraft Factory, Farren was arguably the one who harboured the most genuine passion for aviation, and in later life he became a very accomplished pilot, flying Spitfires, Typhoons, Tempests, and even the Meteor jet aircraft. One of his great strengths was understanding the process of aircraft development: how to coordinate and bring together the disparate elements necessary to ensure the success of a project. Accordingly, he would be selected by O'Gorman to join Chief Engineer Fred Green to project manage the design and construction of the S.E.5a fighter, the second-generation workhorse for the RFC.

Following O'Gorman's resignation in 1916, Fowler insisted Farren continue to work primarily in design, specifically on a seaplane, the C.E.1 ('Coastal Experimental No. 1'), as seen in figure 7.1. The aircraft was given its maiden flight off the River Hamble near Warsash in Hampshire in January 1918, with Grinstead and Lindemann watching on. A second prototype was built that employed a more powerful engine, but the performance of the craft was not

Figure 7.1 William Farren water testing the C.E.1 on the River Hamble, 1918.

impressive enough to tempt the military to issue a contract for mass production, a blow that seemed to affect Farren's enthusiasm for flying. As part of his 'familiarization' to tackle the problem of designing a new seaplane, Farren learned to fly a couple of the existing ones, being cleared solo on the same day, demonstrating his natural aptitude and flexibility as a pilot.

Farren knew that the basic design, construction, and experimental flying needed to be done by people closely connected with a project, just as Busk had realized a few years earlier on the B.E.2 project, so the former was a staunch advocate of allowing pilot-trained academics loose at the controls of developmental aircraft. Farren's contributions at Farnborough during the war, as evidenced by the ACA reports, form an impressive catalogue that reveals the importance of his presence, particularly when it came to putting British aircraft in a position to compete with their ever-improving German competitors.

In 1918, Farren resigned and, with his colleague Green, moved into industry. He filled a niche at Armstrong Whitworth, working on aero engines and engineering issues with cars alongside his primary passion, aircraft design. One of his most notable contributions was his share in the design (again working with Green) of the Siddeley

Figure 7.2 A view of 'The Factory', taken by William Farren in 1915.

Siskin S.R.II: a single-seat fighter aircraft that entered service with the Royal Air Force after the war (figure 7.4).

The Siskin was built around the powerful A.B.C. Dragonfly nine-cylinder, 320-horsepower, radial engine, the first prototype flying in the spring of 1919. Following the merger of Siddeley-Deasy with Armstrong Whitworth, an upgraded version of the aircraft appeared in 1921, powered by the more reliable Jaguar engine. This aircraft was ordered by the military, but with a caveat: the construction had to be all metal. This Air Ministry requirement was not without reason. Seeing the potential for car manufacturers to diversify into aviation and offer huge workshop capacity, this shift in materials specification was necessary: cars were not made from wood and fabric! The first metal version of the Siskin, the Siskin III, took to the air in May 1923 (figure 7.5).

Meanwhile, Farren's time at Armstrong's was short lived, and in 1920 he returned to Cambridge to lecture in engineering and aeronautics and work under Melvill Jones, who was putting together a research team of which Farren would become an important member. Many issues in aircraft design and propulsion needed addressing following the war, particularly drag reduction through streamlining, designing monoplanes that had greater structural integrity, understanding and mitigating against various aerodynamic effects, and improving engine performance.

Continuing a trend started at the Royal Aircraft Factory, Farren lectured using diagrammatic and experimental illustrations,

Figure 7.3 William Farren and his first wife Carol (Warrington) taking tea on the river.

often sketching on a blackboard in isometric projection, preferring graphical rather than analytical expression of results. He drew upon least strain energy methods to analyse redundant structures in biplane wings, and he also investigated the impact on performance of alterations to wing loading and shape.[3] Farren continued to use his practical engineering skills while at Cambridge, designing and constructing a tank that he filled with water, in which he suspended oil drops. He then had a mechanism that could propel a scaled wing section through the liquid so he could see and photographically record the nature of the fluid circulation generated by the aerofoil.

The Cambridge work demanded an on-site wind tunnel, and it was Farren to whom Jones turned to design and build the required apparatus. While tiny in comparison to the main tunnels at the NPL

[3]See p. 161 for details of Italian engineer Alberto Castigliano's least strain energy methods, which were often employed to define the distribution and magnitude of structural loads imposed on aircraft wings.

Figure 7.4 The Siddeley-Deasy S.R.II Siskin.

and the Royal Aircraft Factory, the device, when allied with a short-period flow sensor (also designed by Farren), was able to produce accurate data revealing the nature of wing stall. One of the new findings from the data was that the forces on the wing are different at the stall angle depending on the direction of rotation of the wing as it passes through this angle. In other words, if a wing is stalled by increasing α, the forces experienced at the stall angle differ from those the wing will sense at the same angle as it is un-stalled by decreasing α: a hysteresis effect. The force magnitudes sometimes differed by as much as 50%, which had profound implications for what one needed to consider when conducting aircraft stress analysis applicable to the stall, and indeed when it came to how any particular aircraft might handle during stall initiation and recovery.

Yet another innovation was a device Farren designed and built with Taylor that could be flown on a real aircraft to determine the state of the boundary layer at various points on the wing, thus offering insight into the transition from laminar to turbulent flow in this critical region. This prompted a much clearer understanding of the origins of wing drag.

In addition to being a member of the ARC, notably the chair of the Design Panel of the Aerodynamics Sub-Committee from 1923 to 1925, and spending more time with Armstrong Whitworth as a consultant, Farren was elected FRS in 1945 and became a linchpin

Figure 7.5 A No. 41 Squadron Armstrong Whitworth S.R.IIIa Siskin being serviced with oxygen at Northolt in 1924.

of British aviation. It is therefore remarkable and somewhat mystifying that, despite this status and pedigree, very few people know of him. Aerodynamicist Arnold A. Hall sums up why this might be (Hall 1970, 898):

> He would have been the last to claim any outstanding contributions to any of these spheres in which he worked [design, research, technical administration, and teaching] – and rightly, in a sense that no distinguished aeroplane was wholly his, no breakthrough scientific discovery carries his name, and no new administrative edifice was erected to him. Nonetheless he was a great man, and a man of a kind much needed.

Hall was educated at Clare College, Cambridge, where he obtained a first class honours degree in the Mechanical Sciences Tripos during the 1930s, specializing in aeronautics. A multiple prize winner, he spent World War II at the Royal Aircraft Establishment, eventually becoming its director in the early 1950s. He is regarded as one of the most influential British aerodynamicists of the twentieth century. A family archive exists that contains a number of wonderful wartime

Figure 7.6 Cockpit view: port (left-hand side) and rear (right-hand side).

photographs that belonged to Farren.[4] They offer great insight into the flying aspects of the academics' roles at Farnborough. In-flight vistas from Farren's cockpit (figure 7.6) give a real feel for the nature of the construction of early British aircraft, and the images of him on the ground performing various duties offer a notion of life as a research pilot.

THE ADMIRALTY LINK

The industrialization and rapid expansion of aeronautics in some ways made it more difficult for standards to be imposed upon the design and performance of aircraft. While the civilian sector had free rein in many respects, a tighter grip was kept on what was and was not acceptable for aircraft that needed to operate in a military environment. Central to this control were the directives and recommendations issued by the ACA and the British Admiralty, informed by data and experience derived from the Royal Aircraft Factory and the NPL. The Technical Section at the Admiralty in London had very close links with the research establishments, and it is to the former that we now turn. Who were the people that were chosen to work in the field of aeronautics at the Admiralty? What were their roles, and how did they interact with their counterparts at Farnborough and Teddington? Three specific case studies will now be used to help

[4]Archive details can be found on p. 247.

answer these questions, the main focus being the important contributions of women mathematicians in British aeronautics during World War I.

STRESSING AT THE ADMIRALTY

Data for the causes of aircraft structural failure during the early 1900s are elusive, but there is no doubt that a problem existed. What are now termed 'aeroelastic effects' were not fully known or appreciated in the early days of aviation, and they were undoubtedly responsible for some airborne catastrophes. The understanding of these phenomena was a significant missing piece of the aerodynamic puzzle, and their nature is considered in more depth in the appendix.

The application of mathematics to solve aeronautical engineering problems was also very much in a transitional phase: the established mechanics of solid, ground-based structures were having to be rapidly modified to apply to the more dynamic, subtle, and often unpredictable demands of an aircraft in flight. In terms of aircraft design, it was structural integrity that in many ways was responsible for the preponderance of biplanes. The box-like nature of the double wing, with its struts and wires, helped prevent wing distortion due to aerodynamic forces. It was also a configuration that lent itself to structural analysis that was not dissimilar to that pertaining to something such as a box-girder bridge.

One of the most frightening effects experienced by some early aircraft was that of 'flutter': one of the most common aeroelastic issues, it created oscillations in wing or fuselage structures that could potentially lead to excessive twisting forces. The first study into this effect was published in 1916 in the *Reports and Memoranda (R&M) of the Advisory Committee for Aeronautics* (Fage and Bairstow 1920), and it reported on an investigation that would eventually lead to Cambridge mathematician Robert Frazer pioneering the application of matrix techniques to engineering problems.[5] Cambridge Wrangler Frazer had been employed at the NPL to work on viscous flow problems, but once he was drawn into the flutter

[5]For a discussion of the circumstances prompting this research, see Collar (1978b, 38).

debate by Fage and Bairstow, gaining a better understanding of the effect became a lifelong interest and pursuit.

Aerodynamic loading due to aircraft control inputs during manoeuvring was also an area of aeronautics that needed constant appraisal and attention to ensure airframe structural integrity. An initial deliberation of the ACA in this respect concerned load factor: how much g-force should an aircraft actually be designed to withstand? A 1913–14 technical report of the ACA (Rayleigh 1915, 21) answered the question: 6g. As the war progressed and the roles of aircraft became more diverse, however, common sense prevailed, and the design load maximum for larger machines such as the Handley Page O/400 bomber was reduced to 3g; the weight penalty of greater strength would have been practically prohibitive given the modest horsepower of the available aero engines.

Addressing and solving the issue of aircraft falling apart in flight was not, of course, as simple as imposing a blanket g-force limitation on the airframe: failures always occurred at the weakest part of the structure in any given set of circumstances, and catering for all in-flight scenarios was practically impossible. It thus became an iterative process of education, where knowledge gleaned from destructive and non-destructive testing was combined with lessons learned from actual flight failures. Superimposed on these revelations were theoretical stress calculations that informed subsequent design criteria. In the next section we highlight the contributions to this effort of a trio of influential women mathematicians who were employed at the British Admiralty during the second half of the war.

THE ADMIRALTY IN 1917

In 1917, as the air war raged in Europe, involving an ever-increasing number of aircraft and aircrew, the need to ensure the structural integrity of existing and new aircraft became paramount. In London, Wing Commander Alec Ogilvie led the Technical Section of the Admiralty Air Department, which housed a number of individuals dedicated to addressing aircraft structural issues. Among them were three women: Hilda Hudson, Letitia Chitty, and Beatrice Cave-Browne-Cave. Even allowing for the influence and impact of the war, finding women mathematicians occupying such positions in those days was unusual, to say the least, and the stories that reveal how their appointments came about are certainly of interest.

Militarily, the command and control of aerial assets was centred at the Admiralty, and while General Henderson pondered the amalgamation of the RFC and the Royal Naval Air Service, the important business of ensuring the structural integrity of Britain's military aircraft continued apace. The rapidly changing dynamics of the conflict in terms of air power ensured that the organizational structure of the Admiralty during the latter years of the war was in a state of flux, and with more than 10,000 assorted military personnel and civilian employees concentrated in London it must have been a frantic and challenging environment in which to be applying mathematics.

The focus here is on considering the impact that the three above-named women had on the effort to make early British-designed aircraft as structurally sound as possible. Working in the structures subdivision of the Technical Section, the women were located in offices in the Hotel Cecil, just off the Strand, in the very heart of the capital. Before considering their individual contributions in detail, however, it is worth noting some of the influential characters present in their domain. While it is not a comprehensive reflection of those present, figure 7.7 illustrates where in the hierarchy the women sat in 1917.[6]

As stated, Alec Ogilvie was the controller of the Technical Section. He had an impressive pedigree in aviation: he was the seventh person in Britain to hold a pilot's licence, and he was a pioneering competitor in the early air races. He came third in the 'Gordon Bennett' air race of 1910, flying a Wright Model R aircraft, and he backed this up with a fourth place in the same event a year later. His technical pursuits resulted in the production of an early type of airspeed indicator, and his impressive curriculum vitae would prove to be a future passport into an important leadership role when the Royal Air Force was formed in 1918.

Leonard Bairstow operated as the master overseer, coordinating efforts across the various technical disciplines of structures, aerodynamics, performance, and airscrews (propellers); he appeared to have more direct control over the work designated to Beatrice Cave-Browne-Cave than over that given to the other women mathematicians. Cave-Browne-Cave is known to have been paired up with

[6]Skempton (1970, 466) indicates that there were about 20 people working in the structures office by 1918.

Figure 7.7 The Technical Section of the Admiralty Air Department (1917).

Eleanor Lang, who had graduated with an MA in mathematics from UCL in 1911.[7]

Albert Thurston's status was similar to Bairstow's, but his remit was more on safety and testing. He officially occupied the seat below the controller, but Ogilvie's second-in-command in structures was, in any practical sense, Sutton Pippard (see p. 161). Arthur Berry and Laurence Pritchard were Pippard's henchmen. Berry was the Senior Wrangler in the Mathematical Tripos of 1885 and had remained at Cambridge University as a fellow of King's College and a lecturer in mathematics. Pritchard, born in 1885, also completed his undergraduate mathematics degree at Cambridge. It is likely that the paths of the latter two men crossed in Cambridge during the early 1900s, before they were later reunited at the Admiralty. Berry's prowess in mathematics would have complemented Pippard's engineering insight and experience perfectly. In addition to

[7]In Chitty's words, the two were Bairstow's 'devoted assistants' (Chitty 1966, 67).

his published works in aeronautics, Pritchard would go on to publish a number of detective novels under the pseudonym John Laurence. In one of these, *Murder in the Stratosphere* (Laurence 1938), he alludes to his thoughts on the future of aeroplane technology. He would also become a stalwart of the Royal Aeronautical Society in the three decades following the war.

Then came the women: Hudson, their chef d'équipe; Cave-Browne-Cave; and Chitty, the relative youngster. Hudson arrived directly from a teaching post and Chitty from university, but Cave-Browne-Cave had trodden a more vexed path, working for Karl Pearson in the Department of Applied Statistics at UCL before leaving under somewhat acrimonious circumstances to take up her post at the Admiralty with Bairstow (see Porter 2010, 274).

Pearson had formed a link with Girton College, Cambridge, thereby providing an opportunity for mathematically gifted women to work on statistical analysis at UCL. Beatrice's younger sister Frances, a fellow Girton alumna, had been placed in the third division of the first class of the Mathematical Tripos in 1899, and she was one of the first to take advantage of the UCL opportunity, co-authoring works such as 'On the correlation between the barometric height at stations on the eastern side of the Atlantic' with Pearson (Cave-Browne-Cave and Pearson 1902).

The Bairstow/Cave-Browne-Cave partnership strengthened and flourished both during and after the war. Similarly, Chitty forged a strong working relationship with her mentor, Pippard – a bond that the pair would subsequently exploit in the interwar years.

THE ENVIRONMENT FOR WOMEN

We now have some appreciation of where these women sat in the overall scheme of British aeronautics in 1917, but what brought each of them to London?

Society did not particularly encourage young women to have aspirations in the fields of mathematics and engineering: these were subjects that were very much considered exclusively male domains. Indeed, few schools were willing to teach girls to a sufficiently high standard in mathematics even to enable them to pursue the subject at university, so private tuition was invariably the only practical way forward. For those few who did follow this path, having managed to gain acceptance into a college such as Girton or Newnham, Cambridge, their relative lack of preparedness made them unattractive candidates in the general melee to secure the best tutors. Their coaching was therefore often, although not exclusively, inferior, as was their general status. Indeed, at Cambridge, it was only after 1880 that women had the right to sit the Tripos examinations, after Charlotte Angas Scott blazed the trail. Scott, an alumna of Girton College, went on to become professor of mathematics at Bryn Mawr College in Pennsylvania, and she established a strong conduit between Britain and the United States for aspirational women mathematicians.

Despite Scott's breakthrough, however, it remained impossible for women to appear in the same ranking list as the men because of their non-degree status: women could only be referenced by the position they would have held had they been on the list. The primary aim of the policy not to award degrees to women was simply to deny them voting rights in university elections, thus depriving them of any power in the university's decision making. It was not until 1948 that this disparity was addressed at Cambridge, although Oxford (1920) and London (1878) were more progressive. Future employment prospects were also clearly limited by the complete male dominance of virtually all spheres of mathematics and engineering (other than school teaching, perhaps, which was unsurprisingly a profession in which all three of the women discussed in this section were employed at some point during their lives).

So both the general status of women who had studied university-level mathematics and their realistic chances of a successful career were poor at best. Hardly surprising, then, that the frustration felt in general among many women was beginning to manifest itself in more demonstrable ways, as the Suffragette movement gained momentum. Societal challenges aside, however, what were the technical issues that faced Hudson, Chitty and Cave-Browne-Cave when they arrived at the Admiralty?

HILDA HUDSON

It could be said that Hilda Hudson was born into mathematics. Her father, William, lectured in mathematics at Cambridge and was subsequently appointed professor of mathematics at King's College, London, shortly after Hilda's birth in 1881. Her mother read mathematics at Newnham College, Cambridge, and her elder brother excelled at St John's, Cambridge, achieving the coveted accolade of Senior Wrangler in the Mathematical Tripos of 1898; Hilda's sister would very creditably be ranked alongside the equal-8th Wranglers on the list of 1900. What pressure there was, then, on Hilda to shine when she arrived at Newnham in that same year. Like her family before her, though, she rose to the challenge and held bragging rights over her sister by achieving a mark equivalent to the 7th Wrangler on the list of 1903 following Part I of the Mathematical Tripos examinations. A year later she took Part II of the Tripos and was placed in the third division of the first class.[8]

After leaving Cambridge in the summer of 1904, Hilda spent the winter semester at the University of Berlin, where she attended lectures given by Hermann Amandus Schwarz, Friedrich Schottky, and Edmund Landau. Schottky had succeeded Lazarus Fuchs at the University of Berlin in 1902, and the former's doctoral thesis, published in 1877, was considered an important contribution in the field of conformal mappings, an area of mathematics that was of keen interest to Hudson following a schooling in the topic by Arthur Berry;[9] indeed, much of Schottky's post-doctoral work was concerned with linear transformations. Schwarz, at the university since the departure of Karl Weierstrass in 1882, also had strong links with conformal mappings. Issai Schur, too, was lecturing in Berlin while working on the projective representations of groups during Hudson's stay in the city.

It is likely, then, that Schwarz and his colleagues were major influences in developing Hudson's interest in conformal transformations. The subject would come to dominate her mathematical research, culminating in 1927 with the publication of her comprehensive and well-respected treatise *Cremona Transformations in Plane and Space* (Hudson 1927).

[8]For a detailed description of the development and structure of the Mathematical Tripos, see the book *Masters of Theory* (Warwick 2003).

[9]The problem of representing the surface of a sphere on a flat surface is a typical example of where some sort of projection or 'mapping' has to be performed such that every point on the surface of the sphere appears as a unique point on the flat surface. Such transformations can lead to various properties of the original object – things such as distance or area – being distorted on the map. A transformation that locally preserves angles in such a process is known as a *conformal mapping*: angles between lines on the original object and the equivalent lines on the map are the same. Hudson specialized in a class of mappings known as *Cremona transformations*, named after Luigi Cremona, the Italian mathematician who first used and developed them.

Figure 7.8 Arthur Berry.

One must certainly wonder what forces and influences were at play in fostering Hudson's move to Berlin. She would no doubt have been encouraged by her family, and Berry certainly had relevant contacts and experience, having previously taken a sabbatical at Göttingen University to work with Felix Klein as well as having coached Grace Chisholm at Girton prior to her doctoral work under Klein at Göttingen.[10]

Hudson's return from Germany saw her back at Cambridge: first as a lecturer and later as an associate research fellow. Her letter of application for the fellowship offered by Newnham College in 1910 is most revealing, particularly regarding Berry's involvement. Not only do we learn that Hudson met with Berry on numerous occasions in the early part of that year, but she makes it quite clear that he was the one who suggested the line of research she proffered in her application. Hudson's proposal was entitled 'Birational transformations in three dimensions', and it looked to improve and develop work done previously by Arthur Cayley (1870) and Corrado Segre (1897).[11] Furthermore, it would be Berry to whom Eleanor

[10]Grace Chisholm married English mathematician William Henry Young in 1896 and adopted the surname Chisholm Young. After initially being educated in mathematics at Girton College, Cambridge, she moved to Göttingen University in Germany, where in 1895 she became the first woman in that country to receive a doctorate. (See chapter 6 of Claire Jones's *Femininity, Mathematics and Science, 1880–1914* (2009).)

[11]Details are contained in a letter dated 30 April 1910 that is held in the Newnham College archives.

Sidgwick – the college principal and the recipient of Hudson's application – would turn for advice as to the merit and potential of Hudson's proposed research. Unsurprisingly, Berry was most supportive, and on news of Hudson's subsequent appointment to the post, he remarked to Sidgwick: 'I feel sure you have chosen a very good Fellow!'[12]

Perhaps a defining moment in Hudson's life came in 1912 when she became the first women to deliver a talk at the International Congress of Mathematicians (ICM) (Hudson 1912). It should be noted, however, that Hudson's pioneering achievement would have been assumed by the Italian mathematician Laura Pisati at the ICM meeting in Rome four years earlier had it not been for Pisati's death just prior to the 1908 conference.[13]

A short spell at Bryn Mawr College preceded Hudson's appointment as a lecturer in mathematics at the West Ham Municipal Technical Institute, which coincided with the opening of its Junior Engineering School for Boys. Situated in the East End of London, West Ham Municipal Technical Institute opened in 1900 to provide courses in science, engineering, and art for boys. Secretarial and trade courses for girls were added later, and eventually, in 1913, the Junior Engineering School for Boys was opened. This was the seedling from which the University of East London has now grown.

Hudson resigned from her research fellowship at the end of March 1913, two months before its formal conclusion, to facilitate her move to London. The exact reason behind Hudson's decision to leave her teaching post in 1917 and join the civil service is unclear, but the government had been actively running recruitment drives to draw women into the vacuum created in the traditionally male-dominated professions by conscription, which had been introduced for men in 1916. The Military Service Act of March 1916 imposed conscription on all single men in Britain between the ages of 18 and 41 (with some exemptions, including 'conscientious objectors'); a second Act in May of that year widened the net to include married men.

On joining the civil service, Hudson was immediately drafted into the Admiralty to mentor the group of women that would

[12]This quote is taken from a letter dated 1 June 1910, from Berry to Sidgwick, which is held in the Newnham College archives.

[13]Pisati's paper was read by a male delegate, Roberto Marcolongo, as described in Furinghetti (2008, 533–4).

become an essential cog in the wheel of the Stressing Section of the Structures office.[14] One might conjecture that Berry was likely to have been a broker in this appointment, but clear corroborating evidence for this has not been found.

She was slightly older and more experienced than most of her female colleagues and had the presence and work ethic to set a fine example, soon earning herself the title of sub-section director.[15] She also demonstrated her mathematical flexibility by temporarily casting aside her passion for, and expertise in, geometry to enter the applied world of moments, stresses, and strains. This transition should not be underplayed: these are disparate disciplines within mathematics. In addition to acting as the pivot between the key men in the department (Berry, Pritchard, and Pippard) and the women assigned to assist them, Hudson would individually author two notable pieces of work that were published after the war, as we now describe.

'THE STRENGTH OF LATERALLY LOADED STRUTS'

Hudson's first article, 'The strength of laterally loaded struts' (Hudson 1920b), appeared in *Aeronautical Engineering* in June 1920. Hudson's starting point was the knowledge that the formula often employed for the calculation of the strength of wing struts subjected to both lateral and axial loads could, in certain cases, give incorrect results as the proportion of bending to compression was reduced. Indeed, when the lateral load became very small, the formula became worthless. Hudson was particularly interested in parallel, pin-jointed struts subjected to a uniformly distributed side load in addition to compression, as would be found between the wings of a typical biplane. In this piece of work, she first elaborates on the existing methodology before suggesting her modification.

If a strut is absolutely straight and uniform and is subjected to a perfect axial end load, it will theoretically fail when the end load reaches the smaller of its elastic compressive strength, py$A$, or its Euler failing load, P_E. Elastic compressive strength is the product

[14] A bespectacled Hudson can be seen in figure 1.2 on p. 9: a photograph of some of the members of the Admiralty Air Department. She is sitting in the front row, fifth from the left, next to Alec Ogilvie, who is wearing his military uniform with his pilot brevet attached.

[15] Chitty described Hudson as a 'terrific worker' (Chitty 1966, 67).

of the yield point, py, and the area of cross-section, A. The Euler failing load was derived by Leonhard Euler by applying the calculus of variations to elasticity theory, and his equation, which was published in 1744 (Euler 1744, 267), defines the maximum axial force, P_{E}, that can be applied to a long, slender, straight, homogeneous beam before it buckles:

$$P_{\mathrm{E}} = \frac{(\pi)^2 EI}{(KL)^2},$$

where E is Young's modulus, I is the area moment of inertia, and KL is the effective beam length (K's value reflecting the status of the end points of the beam, fixed or pinned).

In practice, however, it was known that this theoretical value applied only to extremely short or extremely long struts. Thermodynamicist William Rankine, who had eclectic interests in science and engineering, had realized this way before struts had aeronautical applications, and another sagacious – and more contemporary – character who empathized was Richard Southwell, who devised by experiment a measure of 'equivalent eccentricity' that could be introduced into the standard equation as a compensatory factor, δ. Southwell had been schooled at Cambridge by Bertram Hopkinson, and his postgraduate interests centred around elastic stability in structures. He had already published work in this area and had just taken up a post as a lecturer at Trinity when war broke out. Southwell's initial work on struts was published in *The Engineer* in 1912 (Southwell 1912), and he followed this up with further work investigating the failure of metal tubes and formulating a general theory for elastic stability (Southwell 1915a,b,c, 1913).

In 1886, before Southwell made his breakthrough, some of John Perry's lecture notes on the subject of struts had appeared in the weekly technical publication *The Engineer* (Perry and Ayrton 1886). Perry's equation derived therein was Hudson's opening gambit:

$$\mathrm{py} = \frac{PN}{A} + \left(\frac{MN}{Z}\right)\left(\frac{1}{1 - (PN/P_{\mathrm{E}})}\right).$$

She was solely interested in establishing more realistic values for the load factor, N, given values for the yield point of the material, py, the unit compressive load, P, the cross-sectional area, A, the maximum bending moment due to distributive load alone, M, the modulus of strength in the direction of failure, Z, and the Eulerian failing load, P_{E}, discussed earlier. Introducing Southwell's δ

Table 7.1 Hudson's percentage reduction of load factor for continuous beams.

	L/k values								
C	20	40	50	60	70	80	100	120	140
0.1	0.5	1.0	1.5	2.0	3.0	4.0	5.0	6.0	9.0
0.2	1.0	2.5	3.5	4.0	6.0	8.5	16.0	50.0	—
0.3	2.0	4.5	6.0	9.0	12.0	18.0	72.0	—	—
0.4	2.5	6.0	9.0	14.0	23.0	43.0	—	—	—
0.5	3.5	8.0	14.0	23.0	44.0	—	—	—	—
0.6	4.0	11.0	20.0	39.0	63.0	—	—	—	—
0.7	5.0	15.0	31.0	63.0	—	—	—	—	—
0.8	6.0	23.0	50.0	80.0	—	—	—	—	—
0.9	7.0	41.0	77.0	—	—	—	—	—	—

modified Perry's equation to define a reduced load factor, N_1, thus:

$$\mathrm{py} = \frac{PN_1}{A} + \left(\frac{(M+\delta)N_1}{Z}\right)\left(\frac{1}{1-(PN_1/P_{\mathrm{E}})}\right).$$

Despite this modification, there was still a lack of consistency between theoretical calculations and experimental results. The reason for this was that the maximum bending moment due to distributed load alone, M, could be increased or, equally, decreased by any eccentricity present; values for δ derived from experimentation were therefore often erroneous. Hudson wanted to impose a further constraint on load factors to cater for worst-case scenarios, so she introduced three new variables: her safety factor would depend on the quotient of strut length, L, and its least radius of gyration, k, and also upon a measure of the direct stress on the strut, C, where C is defined as the quotient of PN and $\mathrm{py}A$.

Table 7.1 shows Hudson's final corrections, where C is read vertically and L/k horizontally. For example, if $C = 0.6$ and $L/k = 70$, then the required percentage reduction in load factor would be 63%. Rather disappointingly, however, after all her deliberation and reasoned argument, Hudson's conclusion is rather vague: she admits that eccentricity is actually so ill-determined that in reality all her numerical values are rather arbitrary and that all one can really take away from the work is that some allowance should be made, and that any such allowance may need to be appreciable.

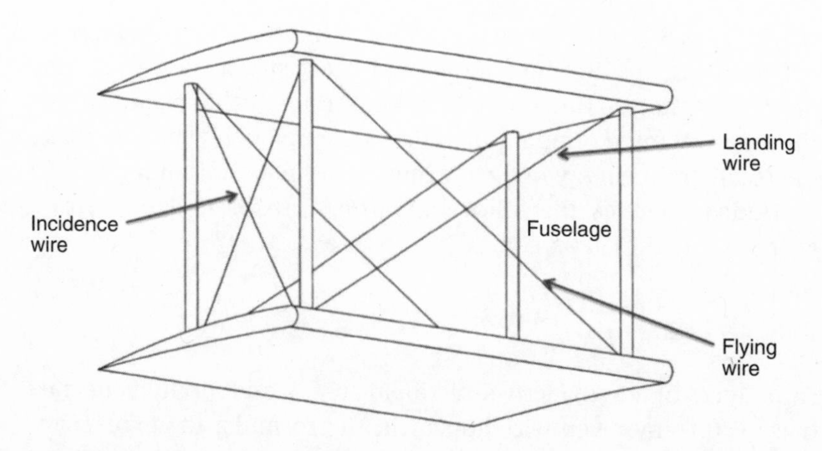

Figure 7.9 Aircraft rigging wires.

'INCIDENCE WIRES'

Her second piece, 'Incidence wires', appeared in *The Aeronautical Journal* in 1920 (Hudson 1920a). In order to be clear about the function of incidence wires, we need to consider three of the general classifications of wires these early aircraft employed. Wires that worked against the wings as they provided normal lift in flight were called 'flying wires', while wires that held the wings up on the ground and supported the wings during landing impact were known as 'landing wires'. The wires that ran diagonally between fore and aft struts were known as 'incidence wires', and they were in place to prevent the wing sections twisting relative to their mountings at the fuselage or relative to each other in the case of biplanes and the later triplanes (figure 7.9). A fourth classification of wires, 'control wires', existed to enable the pilot to move flying control surfaces such as elevators, ailerons, and rudders.

Traditionally, aeroplanes had been treated as braced structures containing a number of redundant members that were not included in any initial strength calculations. In her summary of this paper, however, Hudson emphasizes the serious increase in load on the lower wing of a biplane during a nose dive. This was the justification for insisting on a second approximation that took account of the redistribution of load in such circumstances. Hudson's important work focused on calculating the strains in the incidence wires under various flight modes: level flight, in a nose dive, and in the situation where another wire may have been severed by some mechanical failure or as a result of enemy action. The mathematics of these calculations centres around a method of 'least strain energy', relying on the fact that, as Hudson puts it, 'for strains within elastic limits, the loads in any redundant members take up such values as make the total strain energy of a connected system a minimum'.

Hudson derives the following expression for the total strain energy (Hudson 1920a, 506):

$$\sum \left(\frac{l^3}{2EA} \right) (p_0 + p_1 t_1 + p_2 t_2 + \cdots + p_n t_n)^2.$$

Parameters of length l, cross-sectional area A, and modulus of elasticity E for any given wire appear in the formula; the expression $p_0 + p_1 t_1 + p_2 t_2 + \cdots + p_n t_n$ represents the tension per unit length in any given wire as a function of the load induced in that wire by any external forces when all n redundant wires are slack, p_0, plus the potential load contributions in any given circumstance from each of those redundant wires, $p_1 t_1$, etc. Hudson uses Σ in her original paper in a rather confusing manner to indicate the total strain energy of the system being considered, rather than in the traditional sense in which we would expect to see some sort of summation variable and limits over which the sum is being taken included in the nomenclature.

Partial differentiation of her expression for strain energy with respect to $t_1, t_2, \ldots, t_n$, in turn, produces a set of n partial derivatives that she is justified in setting to zero to define the minimum energy state, and which can then be solved simultaneously:

$$\begin{aligned}
\sum \left(\frac{l^3}{EA} \right) p_1 (p_0 + p_1 t_1 + p_2 t_2 + \cdots + p_n t_n) &= 0, \\
\sum \left(\frac{l^3}{EA} \right) p_2 (p_0 + p_1 t_1 + p_2 t_2 + \cdots + p_n t_n) &= 0, \\
&\vdots \\
\sum \left(\frac{l^3}{EA} \right) p_n (p_0 + p_1 t_1 + p_2 t_2 + \cdots + p_n t_n) &= 0.
\end{aligned}$$

In the remainder of her paper, Hudson takes the reader through various numerical examples that illustrate the application of this technique to different wing configurations. The origin of this method of calculation has an appropriate, if tenuous, link to Hudson herself, and can be traced back to Turin.

CASTIGLIANO

It was the Italian mathematician Alberto Castigliano who, in his 1873 dissertation 'Intorno ai sistemi elastici' ('Concerning elastic systems') (Castigliano 1873), first postulated the method of using partial derivatives of strain energies in calculations such as those being employed by Hudson. He built on previous work completed in this field by the French thermodynamicist Émile Clapeyron in the 1820s and, more contemporarily, by his compatriot Luigi Menabrea.

Menabrea, a professor of mechanics and construction at both the Military Academy of the Kingdom of Sardinia and the University of Turin, achieved posthumous notoriety as the author of 'Notions sur la machine analytique de M Charles Babbage', an account of Charles Babbage's lectures of 1840 at the Turin Academy of Sciences outlining the workings of his 'analytical engine' (Menabrea 1842).[16] Regrettably, however, there ensued a rather bitter priority dispute between Castigliano and his fellow Italian regarding this work since Menabrea's 'energy method' concept lay at the heart of Castigliano's more mature offering. This spat prompted a special meeting of a committee of the Accademia dei Lincei, chaired by Luigi Cremona, whose birational transformations inspired Hudson's defining work.[17] Sadly, Cremona's death coincided with Hudson completing her degree, so he would never witness her post-war homage.

SUTTON PIPPARD

Adding to this mathematically intricate web, it is documented that Hudson's mentor, Sutton Pippard, had drawn great inspiration from Castigliano's work while studying for his degree at Bristol University, so he may well have schooled Hudson in the nuances of the Italian's stress analysis techniques (Skempton 1970, 464).

Pippard had been appointed as technical adviser to the director of the Air Department of the Admiralty in 1915. He soon established himself as a key figure in the field of structural analysis,

[16] Ada Lovelace, daughter of Lord Byron and friend of Babbage, famously translated Menabrea's text (between 1842 and 1843) and added her own set of elaborate comments that she simply called 'Notes' (Lovelace 1843). This additional commentary contains what many identify as the first ever computer program.

[17] Bruno Boley gives a full account of this story in his entry for Castigliano in the *Complete Dictionary of Scientific Biography* (Boley 2008).

and he would go on to become an influential figure at Imperial College, which now holds his personal archive. His key legacy from the war years was to co-author two books with Pritchard: *Handbook of Strength Calculations* (Pippard and Pritchard 1918) and *Aeroplane Structures* (Pippard and Pritchard 1919), with the latter being described in *Nature* as 'an authoritative account of one of the most important aspects of aeroplane design, as well as aeroplane theory ... [it] will no doubt be the standard work on the subject in English for some considerable time' (Brodetsky 1921c). Selig Brodetsky, joint Senior Wrangler in 1908 and author of the review in question, was an influential figure in aerodynamics in Britain during the first half of the twentieth century, establishing himself at the University of Leeds, where he occupied the chair of applied mathematics between 1924 and 1948. His post-war work *The Mechanical Principles of the Aeroplane* (Brodetsky 1921b), however, attracted some rather polarized reviews (Barrow-Green 2014, 108).[18]

Brodetsky's bold prophecy proved insightful, as the work did indeed become the standard text for interested parties throughout the interwar years, appearing in a revised edition as late as 1935. Pippard would also have a significant influence on the second of the women under the spotlight in this story: Letitia Chitty.

LETITIA CHITTY

In 1914, Letitia Chitty was 17 and was busy working with her private tutors in Winchester. Unlike Hudson, she was not surrounded by mathematicians, but her family were academically minded, with a particular affiliation to Balliol College, Oxford. Her father, Herbert Chitty, studied classics at Balliol and eventually became the archivist at Winchester College. He was considered worthy of recognition at the National Portrait Gallery, London, where a print of him playing cards in later life hangs. Her elder brother, Christopher, also attended Balliol before being ordained. A younger sister, Patience, completes the picture. In 1916, Letitia went up to Newnham to read mathematics. Her talent for the subject soon became apparent, and as demand grew in London for competent mathematicians to assist with the war effort, it was agreed that Chitty would be released at

[18]Less controversial was his book on nomography (Brodetsky 1920), which complemented earlier work by Philibert d'Ocagne in France (see footnote 7 on p. 170).

the end of her first year at Cambridge with the promise that she could resume her studies once war was over.

Rumours of Hudson's work at the Admiralty had already filtered back to Newnham via the dons' network, so that was where Chitty wanted to be. By a somewhat convoluted route, the now 20-year-old Chitty presented herself at the Hotel Cecil in August 1917. She was allocated a shared room with two women already working for Hudson: Dorothy Chandler and Mary Hutchison. Soon she encountered Beatrice Cave-Browne-Cave, who was then working alongside Eleanor Lang for Leonard Bairstow, who had himself only recently moved to the Admiralty. Dorothy Chandler graduated with a BA honours degree in mathematics from Royal Holloway College in 1910, having first studied for the 'intermediate arts' exam, which covered Latin, English, and pure and applied mathematics. Mary Hutchison was a product of the ELC, graduating in 1913 with a BSc honours in mathematics. Both Chandler and Hutchison arrived at the Admiralty a week before Chitty.

Quotes, all taken from Chitty's later recollections of her time at the Admiralty, give us some insight into the mathematics and the mathematical methods being employed, and into the general modus operandi of the Stressing Section at that time:[19]

> We relied upon our slide rules and arithmetic in the margins, supported by the theorem of 3 moments and Southwell's curves for struts.

The 'theorem of 3 moments' here refers to Émile Clapeyron's work in the middle part of the nineteenth century that addressed the relationship between the bending moments at three consecutive supports of a horizontal beam (Clapeyron 1858). In its raw state, the theorem's application to aircraft structures is awkward. Arthur Berry, however, managed to adapt it into a more user-friendly form for wing spar calculations, and his Berry functions are now legendary in the world of stress analysis. Berry functions are tabulated values of awkward, trigonometric expressions that appear as coefficients in moment calculations; they are listed, for example, in an appendix to *Handbook of Strength Calculations* (Pippard and

[19]Chitty contributed to the centenary edition of *The Journal of the Royal Aeronautical Society* in 1966 (Chitty 1966, 67–8).

Figure 7.10 Herbert Chitty with Letitia (left) and Patience in 1917.

Pritchard 1918). The user simply has to look up the relevant value in a table using an angle value as the argument. A typical Berry function with argument α might be

$$\phi(\alpha) = \frac{3}{4}\left(\frac{1 - 2\alpha \cot 2\alpha}{\alpha^2}\right).$$

Berry was arguably the strongest mathematician working for the Admiralty at this time, and he, like Pippard, would have had a significant influence on those working with him. Pippard described Berry as 'an exceptional professional mathematician', but he was amused by the latter's lack of exposure to practical engineering. He recalls the occasion when Berry (who would become vice-provost of King's in 1924) enquired about an 'odd instrument' being used in the office: he had apparently never previously encountered a slide rule before (Pippard 1966, 70)! Because it took a more rigorous approach than the alternatives, the 'Berry method' became mandatory for calculating main spar loading for any aircraft wanting Air Ministry approval.

The 'curves' mentioned by Chitty (figure 7.11) illustrate the contemporary preference for mathematical equations to be presented in a more accessible, graphical form.[20] The example shown simply requires knowledge of the length and diameter of the strut in question to glean an approximation for the relevant crippling load.

Another quote from Chitty noted that:

> Each aeroplane, designed but not yet constructed, was received in the form of drawings and assigned to a pair of workers, the one to stress, the other to check the calculations … Lives were at stake!

This observation tells us a great deal about how different things were in the early days of aviation regarding the process of producing new aircraft. Nowadays, stress analysis is an integral and iterative part of the design process from conception onwards; in 1917 it was necessarily more haphazard due to the incessant pressures of the war, although it was certainly not lost on these women that aircrew lives depended on the accuracy of the mathematical calculations they were doing.

Next we have Chitty's confirmation of Hudson's earlier description of the specific modes of flight considered for stress analysis at the Admiralty:

> The general conditions for stressing were normal flight, nose diving, and 'wires cut' one at a time.

[20]Figure 7.11 is an example of an 'intersection nomogram'. This was one of two types of nomogram in use at the time, the other being an 'alignment nomogram', a good example of which can be seen in figure 7.14.

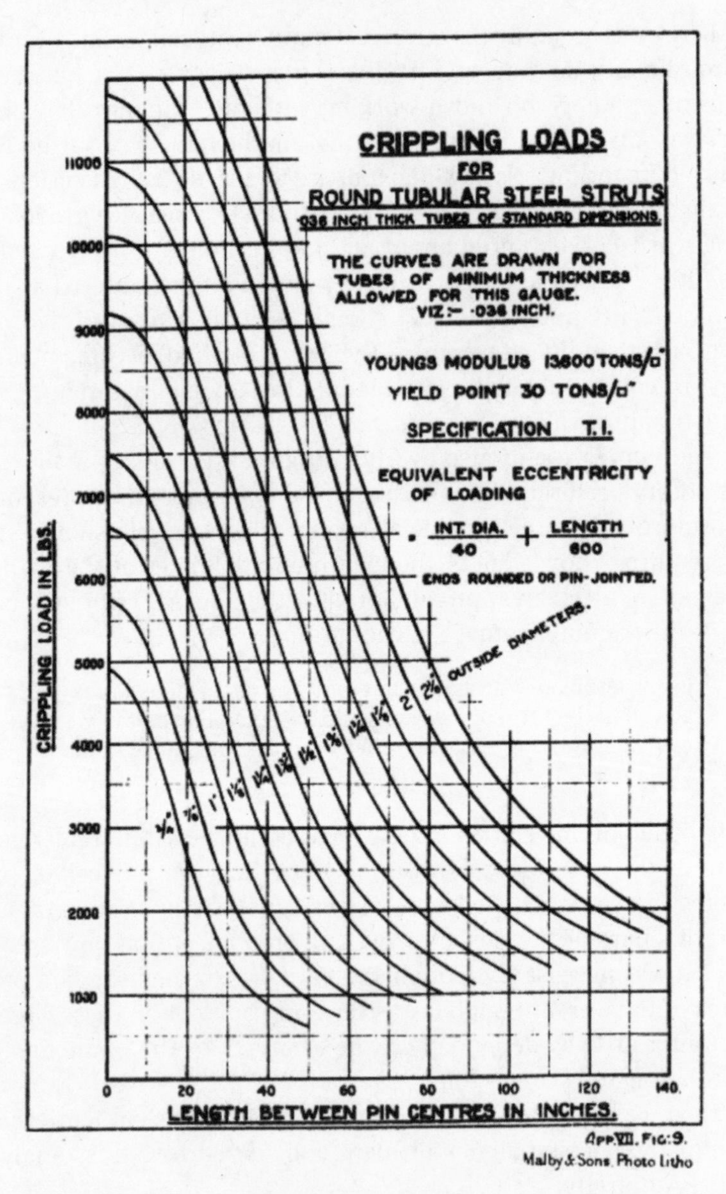

Figure 7.11 Southwell curves for crippling loads.

Despite Pippard and Pritchard grabbing the headlines for running the department and producing the seminal text defining aeronautical strength calculations, a camaraderie that fostered trust and collaboration was clearly present, as the following quote illustrates:

> To begin with Captain Pritchard did most of the teaching; later we taught the newcomers [and] at the end of the War, as a joint effort, we wrote the *Handbook of Strength Calculations*.

The task facing the department as the air battles intensified was huge, so one cannot overemphasize the importance of such teamwork in coping with the burgeoning number of new aircraft being designed and requiring certification.

It is worth emphasizing that sole authorship of ACA technical publications was not usual for women at that time, so any who were granted this honour were exceptional; women were, however, more frequently given credit in a joint capacity. We have already seen that Hudson's individual efforts were published, but in journals rather than as ACA technical reports. Of course, much of the work on stress analysis was classified during the war itself, which explains why some of it did not appear in the public domain until the 1920s and later.

It is clear that Letitia Chitty had been inspired by Sutton Pippard during her time working in the Stressing Section, and there was certainly mutual respect. In fact, Pippard said of Chitty that (Gay 2007, 187):

> She was not just a methodical calculator, but a good theoretician in her own right. She should have been encouraged to publish more on her own, and should, perhaps, have been given a professorship.

Such was Pippard's influence that Chitty returned to Cambridge after the war and immediately transferred from mathematics to engineering, later being placed in the first class in the Mechanical Sciences Tripos, the first woman ever to achieve this distinction. She would later team up with Pippard to undertake stress analysis on all manner of objects, including arches, wheels, dams, and extensible cables.

Pippard was placed in charge of the civil engineering department at Imperial College in 1933, where he was reunited with Chitty when she was appointed his research assistant the following year. And there began their fruitful academic partnership, as witnessed by their many subsequent joint publications spanning the years 1936–60, a list of which can be found in Skempton (1970, 477–8). Prior to this symbiotic collaboration, she also found occasion to work with both Bairstow and Southwell, the latter co-authoring papers with her on hydrodynamic stability (Southwell and Chitty

1930) and stresses in airship hulls (Chitty and Southwell 1931), which demonstrates her remarkable flexibility and gives a clear indication of the lasting impression she must have made during her time at the Admiralty.

BEATRICE CAVE-BROWNE-CAVE

Beatrice Cave-Browne-Cave was educated at home and, like Hilda Hudson, she was blessed with siblings who shared her passion for mathematics.[21] She would eventually go up to Girton in 1895 and come away in 1899 having been placed in the third class in Part II of the Mathematical Tripos – perhaps, with hindsight, a rather modest indication of her mathematical potential.

She immediately took up one of the few options open to female mathematicians and became a teacher at Clapham High School, but it would be an opening at UCL just before war began that would launch Cave-Browne-Cave's career in mathematics. Her sister Frances was employed as a lecturer of mathematics at Girton but had established a concurrent working relationship with Karl Pearson in London. She therefore probably played some part in Beatrice's UCL appointment. Beatrice's initial work was statistical in nature, but as the war intensified, and much to Pearson's chagrin, she took an opportunity to earn more money by working at the Admiralty on aircraft tail loading analysis and the study of aircraft oscillations.

In figures 7.13 and 7.14 we see two of the original diagrams used by Cave-Browne-Cave to calculate the force on a tailplane. The former is her force diagram and the latter depicts her use of an alignment nomogram: a device employed when a single type of calculation had to be done repeatedly. It had been invented by the French engineer Philibert Maurice d'Ocagne in 1884 in response to demand from French engineers for a method to speed up the operations of cut-and-fill necessary to expand France's railway system. D'Ocagne

[21] The interesting history behind the evolution of Beatrice's rather distinctive surname dates back as far as William the Conqueror, who conferred upon two brothers living in the English county of Yorkshire the lordships of South Cave and North Cave in 1069. *The Baronetage of England* (Kimber and Johnson 1771, 355–65) relates the story of the Caves until the mid-1700s, and also illustrates their coat of arms (Kimber and Johnson 1771, A6). Cokayne's *Complete Baronetage* (Cokayne 1902, 93–5) picks up the narrative, explaining the change by Act of Parliament to 'Cave-Browne' in 1752 and by Royal Licence to 'Cave-Browne-Cave' in 1839.

Figure 7.12 Beatrice Cave-Browne-Cave at Girton, 1895.

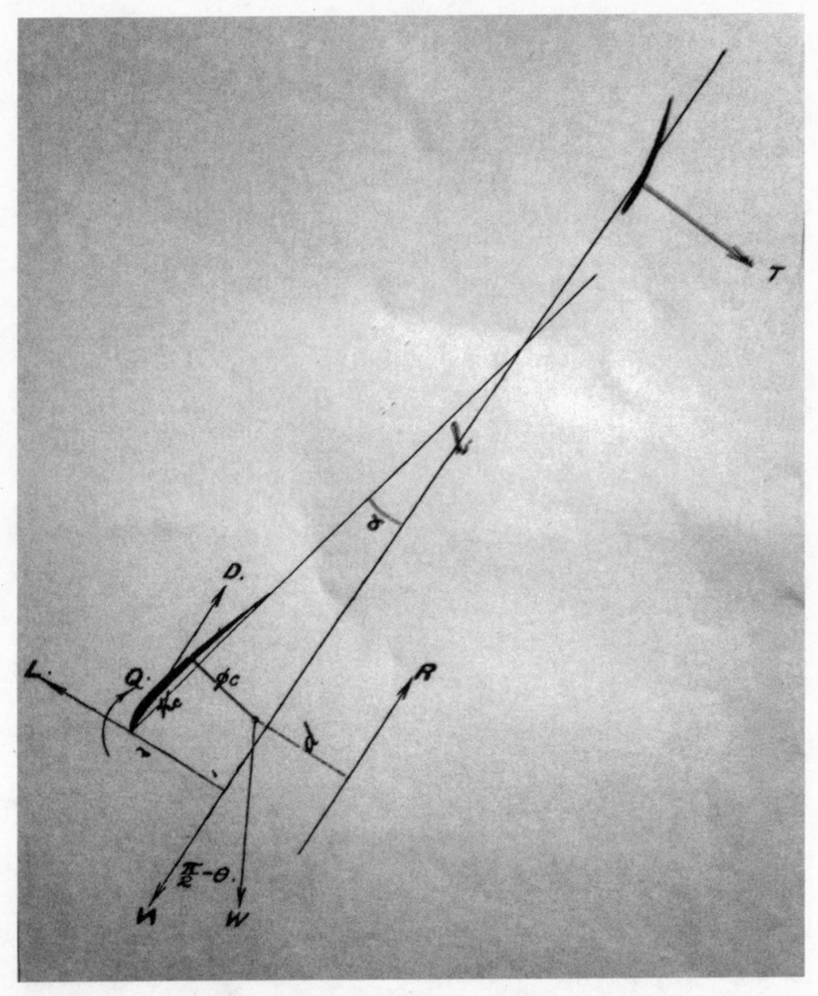

Figure 7.13 Force diagram of an aircraft in a dive.

would go on to write a number of treatises on the study of nomography and other types of graphical calculation methods following his collection and analysis of both intersection and alignment nomograms (Tournès 2016).

To use this particular nomogram she would have started with the input 'weight of machine' on the left (3,000 lbs) and constructed the straight line that connects to the 'main plane chord' (6 ft); this line then defines a point on the reference line through which a second straight line is constructed from 'distance from centre of gravity to tail plane' (15 ft) to intersect the 'tail load' scale, from which

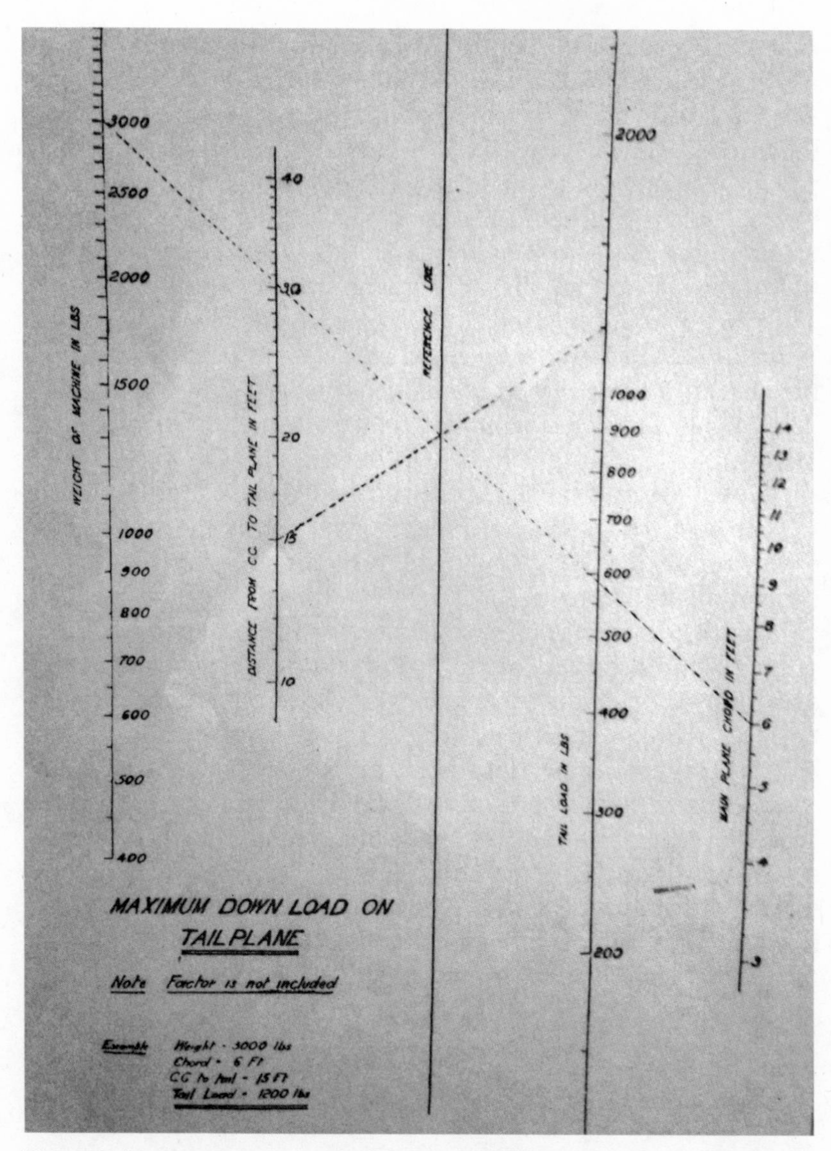

Figure 7.14 Nomogram to calculate the maximum load on a tailplane.

the load on the tail can be read off (1,200 lbs in this case). This is clearly a simple method of calculation for anyone averse to algebra or slide rules, so it would have been welcomed in industry.

Her study of aircraft oscillations was published in an ACA technical report (Cave-Browne-Cave 1922) that demonstrated her sound grasp of the necessary mathematics. As the war ended, Leonard

Bairstow, who was moving to become the first Zaharoff Professor of Aerodynamics at Imperial College, asked Cave-Browne-Cave and Lang to assist him with research into objects moving in viscous fluids; both women were acknowledged in the resulting published papers (Bairstow, Cave, and Lang 1922, 1923). The mathematics here is rather complex; quite how much of it was developed by Bairstow and how much by his assistants is not clear. It is likely that the women would have been employed to check Bairstow's work and also develop pieces of original mathematics delegated to them by Bairstow that he could subsequently incorporate into the general flow of the proofs. Either way, transitioning into the field of fluid dynamics from that of stress analysis shows great mathematical versatility on the part of Cave-Browne-Cave.

As the structures team at the Admiralty moved into top gear to cope with the significant demands imposed by the second half of the war, there was a changing of the guard at Farnborough. The next chapter brings to the fore the roles of two mathematicians who became flight observers: David Pinsent and Hugh Renwick. Much of the former's story is told through his own personal letters and diary entries, while the latter's impact is traced through the flight logbook record of Roderic Hill, the chief test pilot at the Royal Aircraft Factory at the time. Hill's story will also be highlighted. He was a larger-than-life character whose flying skills and empathy with the academic cadre at the Factory made him a central figure in British aeronautics at the time. His contributions were complemented by those of his younger brother, Geoffrey, and both Roderic and Geoffrey were undoubtedly influenced by the mathematical prowess of their father Micaiah, professor of mathematics at UCL.

8

Two Observers, Three Hills

THE OBSERVERS

With no sign of the war ending, the importance of the work being undertaken at Farnborough was obvious to anyone involved in aeronautics, not least those in the RFC. Staffing levels among the academics at Chudleigh had to be maintained, so any departures or fatalities meant that replacements had to be sourced expeditiously. Thus, no sooner had Geoffrey Taylor and Fred Green departed for pastures new than David Pinsent and Hugh Renwick took their places in the Chudleigh mess alongside stalwarts Farren, Lindemann, Grinstead, Aston, Thomson, Wood, and someone we are yet to meet: Hermann Glauert. And so continued the pantheon that was the Chudleigh lot.

Pinsent's involvement at Farnborough was significant. Not only was he able to continue the partially completed work of Taylor and contribute to many other tests, experiments, and pieces of theory, but he also kept a diary and wrote frequently to his relatives and friends. This literature – the personal diaries and letters of a man working at the cutting edge of aeronautical development in Britain during World War I – represents first-hand evidence of what was happening at Farnborough during the final two years of the conflict. Moreover, in the context of this narrative, Pinsent is even more pivotal, as his correspondence and notes stretch back to his days at Cambridge University. We can therefore both learn aeronautical things from him and gain an insight into life as an undergraduate mathematician in the years just prior to the war.

The importance of the information that can be gleaned from this personal archive for aeronautical research cannot be overstated.[1] The diaries have already been extensively used by the

[1]The diaries form part of a family archive held by Pinsent's niece, Anne Keynes,

Finnish philosopher and author Georg von Wright to tell of Ludwig Wittgenstein's youth (von Wright 1990), and reading them with a slightly different bias and intent reveals a wealth of detail about Pinsent himself and his time both at Cambridge and at the Royal Aircraft Factory.

We start by stepping back to 1910, allowing Pinsent's thoughts and observations about topics that are relevant to this narrative to guide us through the decade that followed. In so doing, and while appreciating the intrinsic historical value and merit of the archive, we should not lose sight of the fact that these are the thoughts of a young man who gave his life to pursuing the understanding that has kept generations of aviators safe to this day.

DAVID HUME PINSENT

David Hume Pinsent was born in Edgbaston in Birmingham in 1891. He had a younger brother, Richard, and a younger sister, Hester.[2] He was named after the philosopher David Hume, an ancestor of his, and he attended Marlborough College from 1905 to 1910 before going up to Trinity College, Cambridge, to read mathematics in June 1910. During his time at Trinity he came into contact with numerous individuals who would contribute to the story of early aeronautics: he was, for example, great friends with George Thomson and the Thomson family. On 18 November 1910 he writes of his amusement while attending a lecture with George on the subject of electrostatics: a lecture that was almost entirely concerned with the associated theories of George's father (see the top panel of figure 8.2)!

He was also part of Horace Darwin's circle of friends (see the bottom panel of figure 8.2), and could often be found at Arthur Berry's residence taking lunch and, no doubt, talking mathematics (figure 8.3). He also spent a considerable amount of study time in the company of Hermann Glauert.

Glauert was a central character at Chudleigh but, unlike many of his colleagues, not one who had any burning desire to take to

and were kindly made available for this analysis by Professor Simon Keynes at Trinity College, Cambridge.

[2]Richard Parker Pinsent was an 'exhibitioner' in chemistry at Balliol College, Oxford, prior to joining the army. He was killed in October 1915 while on active duty in France.

Figure 8.1 D.H. Pinsent.

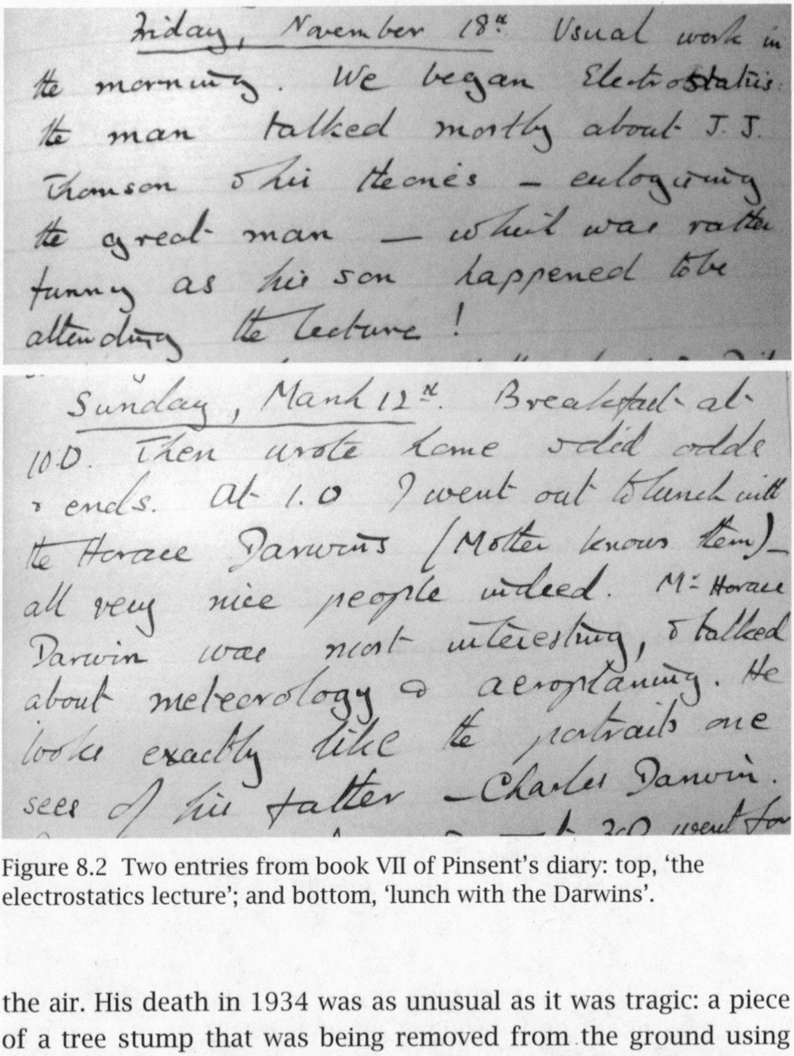

Friday, November 18th. Usual work in the morning. We began Electrostatics: the man talked mostly about J. J. Thomson & his theories – eulogizing the great man – which was rather funny as his son happened to be attending the lecture!

Sunday, March 12th. Breakfast at 10.0. Then wrote home & did odds & ends. At 1.0 I went out to lunch with the Horace Darwins (Mother knows them) – all very nice people indeed. Mr Horace Darwin was most interesting, & talked about meteorology & aeroplaning. He looks exactly like the portraits one sees of his father – Charles Darwin.

Figure 8.2 Two entries from book VII of Pinsent's diary: top, 'the electrostatics lecture'; and bottom, 'lunch with the Darwins'.

the air. His death in 1934 was as unusual as it was tragic: a piece of a tree stump that was being removed from the ground using explosives hit him as he looked on. Glauert's work in aeronautics has been well documented by the aviation historians Jeffrey Ackroyd and Nathan Riley (2011), so we will only go into detail here about one or two points of note. He formalized Lindemann's work on spinning, and by 1919 he was in a position to write a comprehensive account of this work (Glauert 1919). He also extended the theory of longitudinal stability of aircraft by considering the effect of allowing elevators to float freely. Glauert was arguably the most prolific of all those at Chudleigh in terms of his output of papers. In the words of aeronautical engineer Arthur Stephens, Glauert was

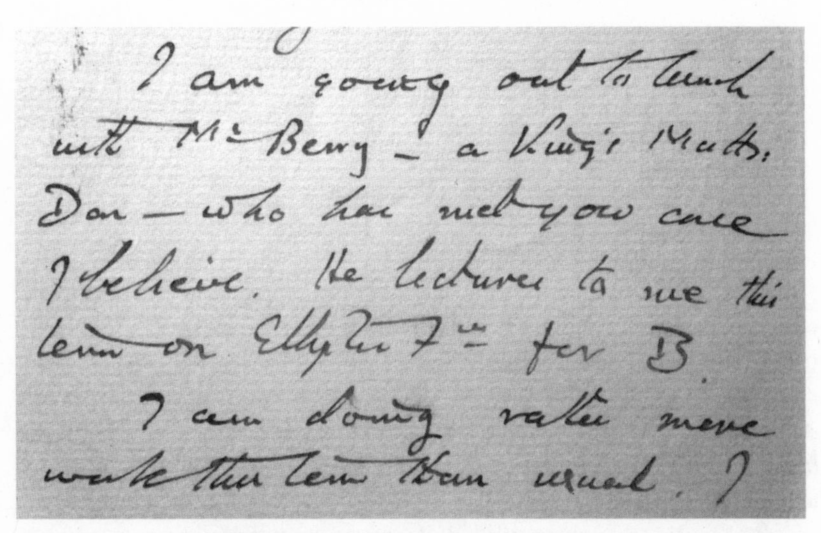

I am going out to lunch with Mr Berry – a King's Maths. Don – who has met you once I believe. He lectures to me this term on Elliptic Fns for B.
I am doing rather more work this term than usual. I

Figure 8.3 A letter from Pinsent to his mother telling of lunch with Berry.

'First and foremost ... a mathematician, with quite unusual powers of verbal exposition' (Stephens 1966, 77).[3]

Pinsent also knew Geoffrey Taylor, the man he would replace years later at the Royal Aircraft Factory, and Richard Glazebrook, the would-be-head of the NPL. And in May 1912, Pinsent met Ludwig Wittgenstein, with whom he formed a strong bond. Such was the closeness of their friendship that the philosopher would dedicate his only published book, *Tractatus logico-philosophicus* (Wittgenstein 1922), to Pinsent. Wittgenstein was on the other side of the conflict once war broke out and was distraught when he received news of Pinsent's death in a flying accident just before hostilities ceased – it was this emotion that prompted the touching dedication. Pinsent's father also provided a lasting memorial to his son, endowing a mathematics prize at Marlborough College: an annual award that is made to this day. The exact circumstances of Pinsent's demise will be discussed shortly.

Book XVIII of Pinsent's diary (spanning September 1913 to March 1914) is where he exposes his passion and talent for mathematics. His idea was to develop a functional calculus dealing with

[3] Arthur Veryan Stephens was a Cambridge graduate who, in 1939, became the first Lawrence Hargrave Professor of Aeronautical Engineering at the University of Sydney in Australia. In 1956, he moved to Belfast to take up the chair of aeronautical engineering at the Queen's University.

a function of the form $F[P(x)]$, where P is a function of x and F is a function of $P(x)$. His aim was to be able to use this calculus to tackle then-unsolvable equations of the form

$$\frac{\mathrm{d}y}{\mathrm{d}x} + y^2 = P(x).$$

He believed this could be solved in the form $\varphi[P(x)]$, where φ is some definite known function of $P(x)$. As soon as P is known, however, he asserts that there is a solution of the form $y = X(x)$, which can be written in the form $y = \varphi[P(x)]$. This detailed mathematical conjecture and reasoning is found in his diary entry of Saturday 8 November 1913: he was clearly taken by the subject and was looking to push its boundaries during any spare moments he had after finishing university.

His talent for mathematics had been formally acknowledged in two stages: a first class honours in Part I of the Mathematical Tripos in 1911, and then, in 1913, a first (and special distinction) in Part II. He found out about his Tripos result on 12 June 1913, via a telegram from his close friend and fellow student M.C. Day.[4] Pinsent was officially 'Wrangler and B-star'.[5]

From his writings, one gets the impression that Pinsent would have enjoyed researching mathematics further in a university environment, but academic posts were few and far between at the time, so he moved his attention towards the army. Unfortunately, though, his frail stature resulted in repeated rejections of his applications. Desperate to make some contribution once war began, he found work at the Ministry of Munitions in London. A letter to his mother dated 8 September 1915 confirms his location and tells of a German Zeppelin raid over the capital:

> Several Zeppelins came right over the Ministry buildings last night and the anti-aircraft gun on the top of Crown Agents office next door was let off at them. They dropped no bombs on us, but made for Woolwich Arsenal which they narrowly missed.

Eventually he did manage to find an opening, and he arrived at the Royal Aircraft Factory on 30 March 1916 to begin work in

[4] The full text of the telegram can be seen on p. 185.

[5] This was during the period of transition from the old 'Wrangler' system of ranking to the new (Warwick 2003, 283–4), and 'B-star' was the highest accolade possible under the revised system.

Figure 8.4 A postcard from World War I depicting a Zeppelin raid over London in 1915.

the machine shop as a turner. His writings indicate that he was completely content and very proud to be working on the shop floor, and had it not been for the fact that George Thomson – his close friend from Trinity who had been at Farnborough for some time – suggested Pinsent's talents were being wasted, who knows how long this underuse of Pinsent's talents would have persisted. A move to H-Department was arranged for 13 November, with Pinsent now recognized as Taylor's natural successor at Chudleigh.

His correspondence tells us much about the schedules being worked, the distribution of labour, the production methods, and, interestingly, the monopoly of the Royal Aircraft Factory in providing certain components to the British aeroplane industry. The wires used for aircraft control and stressing, and the universal joints used to connect the wires together or to the airframe, were all apparently being produced and supplied by Farnborough (figure 8.5). This would clearly have added fuel to the fire in the debate over the Factory's role as a research establishment rather than a manufacturing concern.

So what was Pinsent actually doing during his time at Chudleigh? He clearly slotted straight into life in the academic mess, engaging in discussions about extant technical problems: in an undated letter

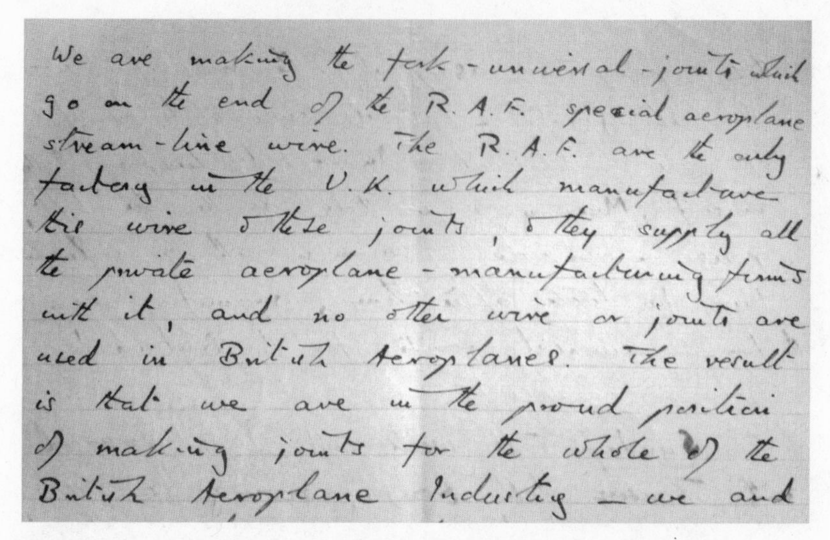

We are making the fork-universal-joints which
go on the end of the R.A.F. special aeroplane
stream-line wire. The R.A.F. are the only
factory in the U.K. which manufacture
this wire & these joints, & they supply all
the private aeroplane-manufacturing firms
with it, and no other wire or joints are
used in British Aeroplanes. The result
is that we are in the proud position
of making joints for the whole of the
British Aeroplane Industry — we and

Figure 8.5 Letter from Pinsent to his mother regarding the Factory's monopoly over the production of wires and universal joints for British aircraft.

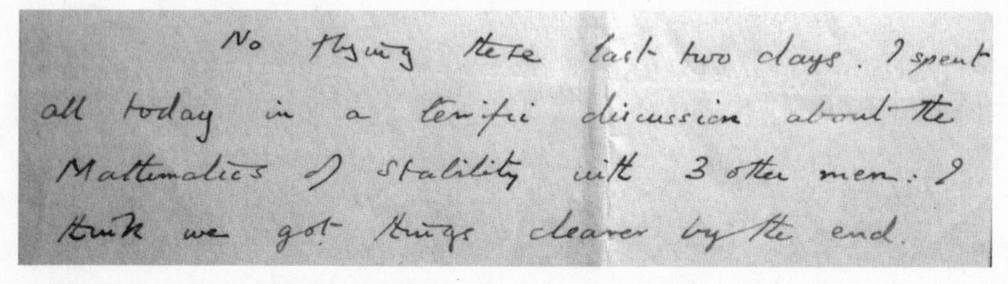

No flying these last two days. I spent
all today in a terrific discussion about the
Mathematics of Stability with 3 other men: I
think we got things clearer by the end.

Figure 8.6 Letter from Pinsent to his mother in which he mentions aircraft stability issues.

to his mother in early 1917, for example, he mentions the 'mathematics of stability' (figure 8.6). Perhaps the most concise summary of his work, however, is given in another letter to Pinsent's mother, dated 14 May 1918, from W. Sydney Smith, the Factory's superintendent,[6] following her son's passing. It reveals much about Pinsent's research role:

> In the aerodynamics department he [Pinsent] soon proved of great value both as a thinker and as a very keen observer. He was engaged on research work on wing sections and the

[6] Sydney Smith took over from Henry Fowler as the superintendent in March 1918, at what had by then become the Royal Aircraft Establishment.

Figure 8.7 Farren (rear) and Pinsent in a D.H.4 aircraft: note the testing equipment stashed next to Pinsent in the front cockpit.

> investigation to determine the difference between the characteristics of models and full scale machines, a matter of great importance in the design of aeroplanes. He was also particularly engaged upon the work of measuring the pressure distribution over the surface of wings in actual flight in continuation of some experiments started by Major Taylor, and in this he did very useful work as he was not only a clever thinker, but a good manipulator in dealing with delicate apparatus. Recently he has been very much interested in the determination of the law of the loss of engine power at height and under different atmospheric conditions, which has always been a difficulty in the comparison of tests made on aeroplanes on different days. He had given much thought to this work, and has recently, with Captain Renwick, proposed a new law of variation. The experiment in which he was engaged at the time of the accident was the measurement of the distribution of pressure on the tail plane of an aeroplane of a type which has recently had several accidents due to the breakage of the tail plane … In the aerodynamic work he was one of the leading members of the department, and it is in his usefulness for original work that his loss will be felt most keenly. It will indeed be very difficult to find another equally suitable to carry on his work.

This can all be corroborated via other sources. Pinsent is mentioned in the 1917–18 technical report of the ACA in 'R&M no. 444', which is entitled 'Exploration of the slipstream velocity in a pusher machine' (Pinsent 1921). He gets joint credit with Hugh Renwick for their study of 'Variation of engine power with height' (Pinsent

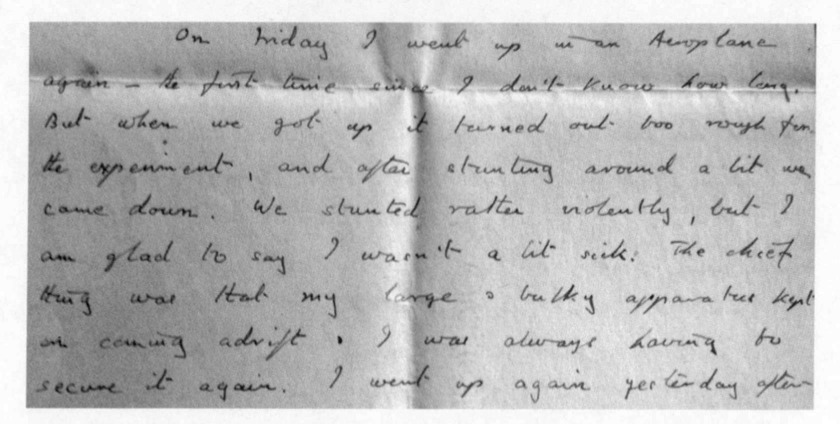

On Friday I went up in an Aeroplane
again – the first time since I don't know how long.
But when we got up it turned out too rough for
the experiment, and after stunting around a bit we
came down. We stunted rather violently, but I
am glad to say I wasn't a bit sick: The chief
thing was that my large & bulky apparatus kept
on coming adrift & I was always having to
secure it again. I went up again yesterday after-

Figure 8.8 Aerobatics and research equipment issues.

and Renwick 1921). The precious pilot logbooks of both the Factory's chief test pilot Roderic Hill and Pinsent's fellow academic aviator Frederick Lindemann also tell tales, in great detail, of Pinsent's activities as a flying observer during flight trials. The picture shown in figure 8.7 highlights a unique aspect of Pinsent's worth in the air: not only were his keen mind and enthusiasm a boon, but his slight frame also meant he was one of the few observers alongside whom bulky test equipment could be carried, something to which he alludes in one of his diary entries (see figure 8.8).

The sad news of Pinsent's death was brought into the public domain by a local newspaper:

> During a routine air test on May 8th 1918 to measure tail loads on a D.H.4 aircraft, Pinsent was observer with pilot Lieutenant Derek Lutyens.[7] Over Frimley in Surrey, the craft was seen to break into 5 pieces in mid-air. Despite a search by 1200 troops, Pinsent's body was not immediately recovered, being discovered some time later in a local canal.[8]

The personal archive of David Pinsent contains so much interesting and worthwhile material that it is difficult to do it justice in the context of a book such as this, so what follows gives merely a

[7] Lutyens was 23 years old and had transferred to the RFC from the Royal Fusiliers in mid-1916. He is buried in the graveyard at Thursley in Surrey, marked by a monument designed by Edwin Lutyens, the renowned architect and a relative of the pilot.

[8] Sydney Smith confirmed in a letter dated 14 April 1918 that Pinsent's body had been found in the canal some distance from the scene of the accident, nearer to Aldershot.

hint of its true historical value and importance. The nature of the insight of Pinsent's writing is illustrated in the following transcript of a letter he wrote to his mother in early 1917. This letter alone tells us when Pinsent moved into Chudleigh House (on 3 January 1917) and that William Farren was in charge of 'messing'. It tells of the movements of important characters, such as Taylor and Green, and we learn of Pinsent's connection with Renwick and that, despite being contemporaries at Cambridge, the two had not met until they both arrived at the Royal Aircraft Factory. We discover the names of all the others living in the house at that time, and we get a perspective on the cost of living: accommodation was clearly not free for these men. We note the duration of a test sortie and the nature of Pinsent's role during that particular flight. We find out that Thomson has been at Farnborough as a flying observer for two years, and we are also given technical information about the policy on aircraft load safety factors. And last but not least, we are given a glimpse of the human aspect of all of this: in a postscript we see a son thoughtfully reassuring his concerned parent that he was going to be safe in his job. If I had to choose the most poignant moment during my research for this book, it would have to be when I read this short note for the first time while knowing Pinsent's eventual fate:

> January 4th, 1917
>
> Dear Mother,
>
> Thank you very much indeed for your letter. I moved along here yesterday evening. Farren, when he asked me to come, told me they had been full up till just recently – when Green and Taylor left. Green has left the factory to look after some private firm, and Taylor has got some job which takes him out of Farnborough. So I have come in – also a man called Renwick – also in H-department – also a Pembroke man and a very nice chap. It is very nice indeed being here – there are Farren, Glauert, Lindemann, Grinstead, Aston, Wood and Thomson and Renwick – besides myself – living here now. Farren really runs the place – and we pay, per month, 1/9 total expenses each: (it comes to about £10 per month): also each man on coming in pays £7, which is returned on leaving – towards Capital Reserve fund.
>
> Last Tuesday I went up again in an Aeroplane – this time as an observer. We were up for $1\frac{3}{4}$ hours on end, and I was writing down.

> P.S. I hope you aren't – please don't worry about me. Of course there is always a certain risk flying – but it is infinitesimal compared with what it would be if I were in the Army. And when one thinks of the numerous people – like Thomson – who have been observing for almost 2 years without a single accident, it must be really very small. The only risk really nowadays is of something breaking whilst in the air – but one must remember that all our machines have 'factor of safety of six'.

A letter three weeks later reveals even more:

> Sunday, Jan 21st, 1917
>
> Thank you all very much for your letters. I could not come home this weekend because, although I am not on duty, there has been a sudden rush job to be done, and I have had to spend all yesterday afternoon and this morning doing Mathematics. It is about that compass business – we are sending in a report to the National Physical Laboratory, and a great deal of my theoretical calculations are going in. I am rather pleased about it, as I think my theory is going to conform pretty well with a lot of experiments that have been done, and I really think it will clear up a lot about compasses that has been hithertoo [sic] obscure. The experiments show quite clearly how a compass behaves badly, and I think my work makes it clear why it does so, and thus how it can be made to behave better. It is rather fine having my work sent to such people as Lanchester, Horace Darwin etc., 'for their information'.

Here we can see that Pinsent was working overtime on the theory to describe compass errors: problems that were resulting in operational disruption and, in some instances, aircrew fatalities and loss of airframes. He is clearly confident and pleased that he has generated some theoretical mathematics that may be of some use in solving the issues, but what jumps out from the letter are a raft of emotions and character traits: excitement at being able to contribute, satisfaction at being able to derive the necessary mathematics, a commitment to putting in the extra time, reverence for his 'seniors', and so forth. In fact, Pinsent's final sentence helps us to really understand the status of those such as Lanchester and Darwin – the established elders in the discipline. (Bairstow would no doubt have been one of Pinsent's 'etc.', given that the work was NPL-bound.)

Moving from his letters to his college diaries, an extract from his entry on 28 May 1913 gives us a first-hand account of the nature of Pinsent's Tripos examinations and his rather candid opinion of the examiners of 1913. His frustration with their entrenched dogma is tangible:

> The papers have not been nice, even the Pure Maths was dirty: there was no Theory of Functions except in the Bookwork paper: the examiners are Forsyth, who is respectable, and Gollop and Stratten, and others like them, who are soulless Mathematicians of the older generation – which perhaps accounts for the lack of questions on such newer subjects as Theory of Functions etc.

His mood was somewhat different a couple of weeks later when the results were posted. This entry paints a picture of an excited young man, pleased with his achievements in mathematics over the previous few years:

> Thursday, June 12th, 1913
>
> I got a telegram from M.C. Day about the Tripos result soon after breakfast – Wrangler and B-star. I am so pleased! I can never bring myself during the exam as to say what I think of what I have done. But I thought in Schedule A that I had just scraped a first: in Schedule B I thought I had done well – but it was very hard to judge – never having been judged in similar subjects before. It is a great relief: now at last I shall be able to read Mathematics as one would any other interesting subject – with no more fear of examinations and consequent modifying of one's course of study. I feel very proud: I have got the top thing all along in Maths – never once falling – an unbroken chain. Scholarship at Marlborough: all three specialists' prizes there: scholarship at Trinity: the same extended in my first year: First in Part I in May – and First B-star in Part II.

Just as Pinsent had replaced Taylor, so Harry Mason Garner, who would become senior scientific officer at Farnborough in 1927, replaced Pinsent at Chudleigh as the war reached its conclusion. Garner was educated at Cambridge, and during the war he found himself seconded from the army to the Admiralty in London, at the request of Arthur Berry, to assist in fuselage stressing and bomb trajectory calculations. Garner's office in Savoy Mansions was destroyed in a Zeppelin attack towards the end of 1916, forcing a move to the Small Drawing Office at the Royal Aircraft Factory. Before replacing Pinsent, Garner had to be assessed as 'fit for purpose' in the role: a test not of the strength of his mathematics but of his stomach, with Roderic Hill taking him up in a B.E.2c aircraft and subjecting him to three successive loops to see how the potential new recruit coped!

HUGH ARCHIBALD RENWICK

Hugh Archibald Renwick is one of the lesser-known mathematicians to grace Chudleigh House during World War I. Raised in Stirling, Scotland, Hugh was one of four children: he had an older brother, William Somerville, a younger brother, Tom, and a sister, Bethea. Tom was a lieutenant in the Rifle Brigade and met his death in action on the front in April 1915, while William, who saw service as a lieutenant in the Royal Air Force, survived the war and went on to inherit the Renwick family fortune, much of which he invested in the sports car manufacturer Aston Martin.

Hugh Renwick's early education was at Cargilfield in Edinburgh, but he travelled south of the border to Pembroke College, Cambridge, for his higher education, eventually taking the Mechanical Sciences Tripos in 1912. He joined the 5th Battalion of the South Wales Borderers at the outbreak of war, and in July 1915 he found himself on the battlefield in France. There, he had a remarkably close shave with death in October, with an enemy bullet passing straight through his upper body, just missing his heart. He made a brief reappearance in France in June 1916 following recuperation at home, but by this time the net to identify mathematically minded engineers to bolster efforts at the Royal Aircraft Factory was being cast far and wide. Accordingly, he soon found himself moving into Chudleigh House, his arrival coinciding with that of Pinsent.

Renwick is known to have worked closely with Pinsent and, as we have seen, the two produced useful information about both the variation of an aircraft's engine power with height and the effect of atmospheric conditions on engine performance, as mentioned in Smith's letter to Pinsent's mother. Like Pinsent, Renwick took to the air as an observer rather than a pilot. Another similarity is that much of Renwick's work is also captured in the flying logbooks of Roderic Hill. The exploits of Hill and Renwick are summarized in table 8.1, which is a consolidated list, taken from Hill's records, of all the test flights the two men operated together.

To the untrained eye this sort of archive material may, at first glance, appear of little historical value, but in fact these entries

Table 8.1 Hugh Renwick's flights with Roderic Hill.

Date	Airframe	Time (mins)	Height (feet)	Hill's notes
15-10-17	RT1	45		'Capt Renwick doing the Speed Courses.'
13-12-17	NE1 (3973)	15	3,000	'Capt Renwick testing rudder forces on a new fin. Did a loop, very good!'
01-02-18	RE9 (3542)	25	4,000	'Capt Renwick Sideslipping experiments but aircraft trimmed badly.'
01-02-18	RE9 (3911)	30	4,000	'Capt Renwick Sideslipping experiments aeroplane trimmed well.'
19-02-18	RE9 (A3542)	10	2,000	'Capt Renwick Test of dihedral. Still badly right wing down.'
19-02-18	RE9 (A3911)	85	4,000	'Capt Renwick Sideslipping experiments. Landed at Hamble. Had tea with Farren and Taylor. Got lost in gloom on way home. Renwick spotted Hindhead and we found the way back.'
16-03-18	RE9 (A3542)	45	5,000	'Sideslipping with Renwick.'
26-04-18	DH6 (B2963)	20		'Test with Renwick.'
28-04-18	BE2E (1793)	20	3,500	'Renwick read pitot in spin.'
17-06-18	RE8 (A8561)			'Sunbeam maori engine — appalling except all out.'

are a gold mine of information that reveal a great deal about the nature of test flying at Farnborough during World War I. Test pilots were evidently not paired up with observers for extended periods, and often more than one test sortie would be flown in a single day. Pilots were clearly qualified to fly many different types and variants of aircraft, and both pilots and observers were involved in multiple and concurrent projects. We are told that aircraft of the same type had individual peculiarities in terms of their flying characteristics, that observers were used in spinning tests, and that sorties tended to be relatively brief in duration. Sideslipping tests were carried out at 4,000–5,000 feet, possibly to allow for a spin recovery if a stall happened with the rudder hard over.[9] Some logbook entries confirm that pilots had time to enjoy some aerobatics on certain sorties. Aerobatics in the sense used here would have meant anything from a steep turn to a full loop. The manoeuvres attempted would have depended on the performance characteristics of the aircraft being flown, its structural limitations, its control systems, and the bravery of the pilot concerned.

The benefit of a pilot having an observer on board to assist with navigation during cross-country sectors is apparent. Aircraft were being used as a means of transport to facilitate meetings between the various parties involved in aeronautics, and pilots were perhaps not always as diligent as they may have been in completing the required flight record in their logbook after a sortie! In the context of this narrative, Hill's logbook for the period 1917–18 proved vital in helping decipher Renwick's specific contributions and roles at Farnborough, particularly given that other sources of such detail are scant.

On 19 August 1918, Renwick had been rostered to fly in an R.E.8 aircraft (airframe number A4205) as an observer on a test flight with pilot Captain Oswald Horsley. The two were operating near the army garrison of Arborfield when disaster struck. The moment was captured by a number of eyewitnesses, and their accounts were given five days later, on 24 August, in the local newspaper, *The Reading Mercury*:

> Private G. Kew A.S.C., Arborfield, said he saw the machine come down. At first it was going very smoothly. Looking up again he saw a large cloud of vapour come out of the back part. A portion of the wing on the left-hand side seemed to come away. There was a kind of explosion, and the machine seemed to lift a little. He could hear the breaking of the machine, which went spinning sideways and then came to earth. Both the officers were dead. Sidney J. Gough, a foreman at the

[9] Sideslipping is a manoeuvre induced by the pilot of an aircraft by inputting a yawing force using the rudder while simultaneously applying a compensating rolling force using ailerons in the opposite sense to maintain aircraft heading. This state of affairs results in a higher drag on the airframe, and it is a common strategy employed by pilots when height needs to be lost quickly while at low speed. A danger is that should a stall occur while sideslipping, the applied rudder may encourage the aircraft to enter a spin.

Figure 8.9 An R.E.8 aircraft similar to the one in which Hugh Renwick lost his life.

> Remount Depot, Arborfield, corroborated.[10] He should think the machine fell from a height of a little over 2,000 feet. One large piece of the wing left the machine – the others were small.

A board of inquiry was hastily convened following the incident, with its main witness being Roderic Hill. Hill reported that he had flown the machine in question on the Friday before the accident under exactly the same conditions as the deceased officers did on the day they died, a Monday. Two adjustments had been required and these had been made. Hill personally carried out a subsequent

[10]The Arborfield garrison was established in 1904 as the Remount Depot, which supplied the military with horses for both operational and ceremonial purposes. It was operated by the Army Remount Service. The depot operated throughout World War I, closing in 1937.

flight test and then gave Captain Horsley the instruction to repeat the test. Everything had been in working order, and the machine had remained in the hangar between the Friday and Monday.

Hill felt that the minor modifications had no bearing on the cause of the accident. Hill had driven out to the scene of the wreckage immediately after news had reached Farnborough, arriving there at 12:30 P.M. He described the wreck as so complete that it was difficult to ascertain what had been broken in the air versus what disintegrated in the crash itself. He confirmed to the inquiry that if the wing had detached as described in the eyewitness accounts, the machine would have been uncontrollable. There was no sign of any explosion or fire. Hill's conclusion was that the wing must have failed, causing the machine to become unmanageable. Furthermore, the pilot was not at fault, and no reasonable inspection could have prevented the accident. Following Hill's analysis, the unique verdict of 'Accidental death through falling through the air in an aeroplane, owing to the failure of the wing structure' was returned. Renwick was buried in Farnborough Cemetery, grave number K.39.

Renwick's story epitomizes the human qualities I want this narrative to highlight about the flying mathematicians. He was well educated in mathematics, and he was keen to volunteer to do whatever was asked of him for his country. Even a bullet through the chest did not shake his resolve. He subsequently and unselfishly put himself at further risk in the cockpits of aircraft that he knew were not guaranteed to be safe, ultimately giving his life in the pursuit of knowledge. People like Hugh Renwick are the unsung heroes of aviation.

THREE HILLS

FATHER

In the year in which Edward Routh was appointed as an assistant tutor at Peterhouse, Micaiah Hill came into the world. He survived the perils of the Indian Mutiny as a young child and was then educated at Blackheath School for the Sons of Missionaries before arriving at UCL in 1872 to study mathematics.

His academic progress was rapid and to some extent mirrored Routh's, with the former obtaining his BA degree in 1874 while enrolled at the latter's alma mater in Cambridge. Indeed, we know that Routh actually taught Hill (see Warwick 2003, 516). Hill

Figure 8.10 Micaiah Hill.

was 4th Wrangler and Smith's Prizeman in 1879, accolades that earned him the chair of mathematics at Mason College in Birmingham in 1880, following a short stint at UCL assisting in its mathematics department.

Mason College, which was established in 1875, was the precursor to what is now the University of Birmingham, its new status being conferred in 1900. As well as producing two British prime ministers in Stanley Baldwin and Neville Chamberlain, the college was also attended by two others who are mentioned in this story: Francis Aston and Henry Fowler. Remarkably, Hill was only 24 years of age when he took this senior post at Mason College.

London would soon beckon again, however, and in 1884 he replaced Richard Charles Rowe as the chair of mathematics at UCL; Rowe had been 3rd Wrangler on the Tripos list of 1877. Two of Hill's contemporaries at the university were Alexander Kennedy, a well-established and respected professor of engineering,[11] and Karl Pearson, appointed to the chair of applied mathematics to coincide with Hill's arrival.

During the last decade or two of the nineteenth century, Hill devoted much of his research time to solving problems in hydrodynamics (Hill 1884). This was clearly a field of great interest to anyone attempting to understand the nature of airflow over a wing of any description, and on this basis we might therefore ascribe to Hill a tenuous connection to the early study and development of aerodynamics. He was also involved in the dialogue between the Royal Aircraft Factory's propeller development team and the mathematicians at UCL who were seconded to assist with the more demanding

[11]Alexander Blackie William Kennedy was a man of eclectic talents encompassing nearly all genres of engineering. His competence and versatility secured him an early engineering professorship at UCL in 1874, and he used this position to leverage support and funding for the construction of the first engineering laboratory in Britain, which was inaugurated in 1878. The mantra here was that students should experience engineering rather than simply watch a 'master' conducting experiments. It was Kennedy's foresight and innovation that brought about the National Physical Laboratory and the Department of Scientific and Industrial Research. Kennedy has further connections to the material in this book, too: he designed the steel structure that supports the Hotel Cecil (see p. 149). His links with the military began in earnest when he was employed on the navy's Dreadnought programme in 1900, and when war broke out in 1914 he took up key positions on influential committees that were responsible for, among other things, gunsights, ordnance, and anti-aircraft weaponry. More about his work in these areas can be found in June Barrow-Green's contribution to *Cambridge Mathematicians' Responses to the First World War* (Barrow-Green 2014, 90).

Figure 8.11 Roderic Hill.

mathematical aspects of the associated theory and carry out the numerical computation of various complex expressions. A much stronger and direct link to aviation, however, would come a number of years later in the form of Micaiah Hill's two sons.

SONS

It was during Micaiah Hill's tenure as professor of mathematics at UCL that his two sons were born. Geoffrey was the younger of the pair, receiving his mathematical education at UCL before starting an apprenticeship at the Royal Aircraft Factory in 1914. The war prompted him to join the RFC in 1915, and he saw action with 29 Squadron in France. Elder brother Roderic, who had originally chosen a career as an architect, also found himself engaged in aerial combat in France until a fateful, and fatal, accident brought him to Farnborough, aged 23. On 28 January 1917, Farnborough's chief pilot, Frank Goodden, was testing the prototype S.E.5 aircraft (airframe A4562) when a wing detached in flight resulting in a crash and his immediate death. This incident was confirmed in a rather matter-of-fact way by Lindemann in a letter written to his father on the very same day. Lindemann wrote: 'Our head pilot got killed today. Nobody can make out quite what happened; probably he tried to loop too quickly and broke the machine.'

Figure 8.12 Frank Goodden.

Goodden was much revered, and the circumstances of the accident cast great doubt on the integrity of the RFC's latest fighter, particularly among the other test pilots. The test pilot cadre at the Royal Aircraft Factory prior to Hill's arrival, excluding the mathematicians, comprised F.W. Goodden, G. de Havilland, N. Spratt, A. Bush, R. Kemp, W. Stutt, and S.C. Winfield-Smith.

Roderic Hill was hurriedly brought back from the front to replace Goodden, and his first task – shortly after he had managed to successfully crash-land a stricken R.E.8 on the local parade square – was to restore faith in the S.E.5, which he did by putting it through its paces over the airfield in full view of all concerned. Hill officially took command of experimental flying at Farnborough on 13 February 1917 having served operationally on 60 Squadron during the Battle of the Somme. He remained in post until April 1923, at which point he began an illustrious career in the Royal Air Force.

It is likely that Roderic had some influence in getting his sibling a post on the test flight at the Royal Aircraft Factory to help rectify the poor spinning characteristics being exhibited by the S.E.5a, and the brothers would eventually fly together to conduct various airborne tests. The perils of spinning were nothing, though, compared with some of the other flying antics with which Roderic Hill would engage.

Perhaps Roderic's bravest exploit was to deliberately fly a modified F.E.2b aircraft into the tethering wire of a barrage balloon at

Figure 8.13 Test pilots Stutt, Spratt, and Winfield-Smith, 1917.

1,000 feet to study the effect of the wire impacting the aircraft wing. He was, unsurprisingly, awarded the Air Force Cross for this incredibly dangerous manoeuvre, which was witnessed from the ground by Lindemann, who had done the mathematics to reassure the pilot prior to the test that he had some hope of survival. The event is recorded in reports H.785 and H.785a in rather modest terms.[12] The reports' titles are somewhat understated – 'The results of a practical trial of a protected aeroplane flying into a kite balloon wire' – and they end with a rather nonchalant conclusion: 'Provided the enemy use thin cables, it appears it should be possible to fly through these balloon barrages without injury.'

This was such a crazy thing to attempt in an aeroplane that it warrants further comment. The technique dreamed up by Lindemann was to employ a robust, V-shaped metal guard with a semi-angle of 60 degrees, fixed in front of the aircraft. The one-off test was carried out by Hill on 27 May 1918 at 5 A.M. at Orford Ness, with the pilot deliberately flying into a cable with a 2.8 mm diameter under 350 pounds of lift from a balloon sitting at 3,700 feet. The F.E.2b was protected with the V-shaped metal guard described above, and the wings were armoured for 6 feet from their tips with sheet iron.

Hill's aircraft hit the cable at 55 miles per hour, and the report describes the predictable result:

> The cable slid down the guard wire as intended to within about 5ft from the wing tip. Here the part of the cable below the guard wire touched the leading edge of the lower wing and

[12]The reports were dated 28 April 1918 and 11 June 1918, respectively.

Figure 8.14 S.E.5a aircraft of No. 32 Squadron RAF in 1918 (notice that the photograph's negative was clearly censored at the time to remove the nearest aircraft's serial number).

> after sliding along it for a short distance caught between abutting ends of the sheet-iron armouring which should have overlapped. This slewed the machine round violently and caused the nose to go down at which point the cable fortunately tore off the armouring and the light wooden leading edge and freed the machine. The machine spun around for about one turn and was then pulled out by the pilot and landed without further damage.

Despite Hill's near-death experience, he remained friends with Lindemann after the war. Hill was secretary of the Oxford branch of the Royal Aeronautical Society during the 1930s, and he was frequently in contact with Lindemann, then a professor at Oxford. Laurence Pritchard, national secretary of the Royal Aeronautical Society at the time, also kept in touch through the activities of this aeronautical body.

The collection of Roderic Hill's flying logbooks held in the archives at the National Aerospace Library at Farnborough is a great legacy. They offer a complete record of his flying career and give great insight into the nature of test flying in those early days of fixed-wing, powered flight. By 1927, Hill had flown an incredible 85 different types of aircraft. As we saw with Renwick's work, the

Figure 8.15 The F.E.2b.

logbooks also allude to others who were involved in the business of the Royal Aircraft Factory: they show Hill flying with the likes of Thomson and Pinsent, and with his own younger brother.

Among the pilot logbook collections encountered during the research for this book, Roderic Hill's is by far the most extensive and the most comprehensive. His flying records tell the story of a pioneering aviator who was at the forefront of experimental and operational flying for a period bracketed by, and including, the two world wars. It is a pity that this collection has not received greater attention because it has so much to offer. Here is a transcript of just one page from one of Hill's logbooks (in relation to 22–30 November 1917), which illustrates the nature of the entries he made during 1917 while he was chief test pilot at the Royal Aircraft Factory:

> Test of aeroplane and engine. Landed on Laffan's Plain.
>
> FARNBOROUGH–STOCKBRIDGE & back. Test of B&B carburettors. Landed and saw Major Strugnall.
>
> Measurement of rudder forces. Came down through clouds S. of HOG's BACK. Made pancake landing and touched a wing tip.
>
> Time for week ending 23.11.17 .... 1 hr 50 mins
>
> Total time as Pilot .... 249 hrs 22 mins
>
> Engine test. Looped & rolled. Tried rolling on top of a loop rather unsuccessfully.

Figure 8.16 Roderic Hill after joining the RFC from the 12th Northumberland Fusiliers in 1916 (note the regimental insignia on the epaulette of his jacket).

> Experiments of variable pitch propellers. Did 5 loops …
>
> Test of B&B carburettors. Engine very dead. Took a long time to pick up after a spin.
>
> Test of comfortable seat …… Did rolls to right and at last managed to do two rolls to left. Came out slightly nose down. Test of Esinar Gun mount aeroplane. Wind very strong. Climbed through clouds at 4000'. Wind strong [illegible text].
>
> Did partial climbs from 9000–10000. Test of 6' chord wings.
>
> Test of Lutning bubble. Climbed up through clouds and let Wood fly the aeroplane.
>
> Test of B&B carburettors. Landed twice on Laffan's Plain.
>
> Test of Daimler "40" conversion on BE12.
>
> Time for week ending 30.11.17 ……… 4 hr 50 mins.
>
> Total time as Pilot ……… 254 hrs 12 mins.

The nonchalance of the throwaway lines is startling: 'Came down through clouds S. of HOG's BACK. Made pancake landing and touched a wing tip.'[13] Nowadays, a wheels-up landing and a wing-tip strike would be major events, but Hill makes them sound like completely routine occurrences: 'Tried rolling on top of a loop rather unsuccessfully.' In today's vernacular, this manoeuvre is called a 'roll off the top'. The aircraft reaches the apex of a loop, at which point it is inverted and at very low speed. Mishandled, any attempt to aileron roll at this point can easily result in a stall or a spin – and Hill clearly found this out: 'Engine very dead. Took a long time to pick up after a spin.' The last thing a pilot needs after recovering from a spin, doubtless disoriented and at very low altitude, is an engine that refuses to start!

In this archive we have a complete record of how the theoretical work of the mathematicians and engineers was integrated with the flying test programme. The nature of the tests, the names of those who conducted them, their frequency, their locations, their success or failure, the quirks of specific aircraft types, aircraft performance characteristics, and flying metrics – it is all here. More than this, though, the logbook entries tell us a great deal about the man himself: his composure, his bravery, his flying skill, his commitment, his sense of humour, and his versatility all shine through. Hill undoubtedly had the 'right stuff', and in spades.

[13] The Hog's Back is a distinctive ridge line in the countryside to the south of Farnborough.

In addition to his time in the air with the various academics when they were acting as observers, Roderic Hill apparently went out of his way to spend many hours with them on the ground, particularly with Thomson, Stevens, and Southwell. This showed an unusual level of tolerance for the academics, who were often avoided by pilots. As Hill's daughter Prudence put it (Hill 1962, 56):

> With his cultured and liberal background, it did not occur to him that the scientist was a creature to be distrusted; someone who, bearded and spectacled, pursued hobbies darkly in the corner of the Factory and was most popular when he did not obtrude himself upon the pilot, that man whose life depended on his own flying skill and not on paper calculations.

One might therefore say that Roderic Hill straddled the gap between academic and practical aeronautics. He was a man who was as prepared to embrace the contributions of the academics as he was to risk his life finding out if the implications of their calculations bore any resemblance to reality. And he could call himself an aeronautical engineer: while he was unqualified in the academic sense, he was extremely experienced in the experimental aspect of the discipline – akin, in many ways, to the wise and invaluable laboratory assistants found in many school chemistry departments, who might not hold a chemistry degree but who play an essential role in setting up practical and accessible demonstrations of intangible theory. Roderic Hill died in 1954 having played a key role in the defence of Britain's airspace during World War II, and, more latterly, having been rector of Imperial College, London.

Like John Dunne (see p. 38), Roderic Hill's brother Geoffrey was particularly interested in aircraft designs that incorporated the feature of swept main wings to help with longitudinal stability; in fact, he took this a stage further and considered the use of asymmetric variable sweep as a means of inducing roll. Rather than having a wing-warping system or conventional ailerons, Hill's notion was that by asymmetrically varying the sweep at the wing tips, he could control an aircraft in roll.[14] In the 1930s, he would take this concept a stage further, proposing that the whole wing should be moveable in sweep. The aircraft design world was not quite ready for such a

[14]A patent (US Patent 1200098) filed by E.F. Gallaudet in the United States in 1914, and granted in 1916, indicates that Hill was not the first person to consider this mechanism for roll control.

radical concept, but the legendary Barnes Wallis was perhaps one of the few in Britain who understood the potential benefits and attempted to study and advance the idea. Of course, many aircraft employ the 'swing wing' in the modern era, to milk aerodynamic benefits at both low and high speeds: the General Dynamics F111 and the Panavia Tornado are typical examples.

THE STATE OF AERONAUTICS IN 1920

Immediately after the war, two books bearing the title *Applied Aerodynamics* were published. Both offered a summary of what was then known about aerodynamics. The books' authors were both well-qualified and credible candidates for producing such a record, but they brought different perspectives to the task at hand. George Thomson was an academic with experience of experimental flying. Leonard Bairstow, on the other hand, was a wind tunnel expert, applying his London-based mathematical education to the analysis of scale models.

Both books cover similar ground but in very different ways. Thomson gives an account that is accessible to someone with a reasonable grounding in mathematics, whereas Bairstow's treatment of the subject is far more demanding of the reader, requiring a far higher level of mathematical know-how to comprehend. These books represent a post-war watershed. Two authors who had been intrinsically embroiled in the frantic quest to advance aeronautics in Britain had gathered their thoughts and drawn upon their memories, their colleagues, and their privileged access to technical documents to produce invaluable snapshots of the field up to the turn of 1919.

By the end of the war, Thomson had returned to Cambridge to teach mathematics, and it was there, in what must have been the relative peace of Corpus Christi College, that he wrote his version of *Applied Aerodynamics* (Thomson 1919). It is rather fitting that the man responsible for enticing Thomson to Farnborough, Mervyn O'Gorman, agreed to write the book's preface, and his opening words are very telling (Thomson 1919, p. v):

> Aeronautical engineering is now a very large subject. It includes mechanical developments which are as much specialized as shipbuilding or locomotive designs, and in addition it has a scientific aspect of its very own. Some of its branches are appropriately to be described as 'obstruse', because of the

> unusual forms of attack which the problems have called for experimentally, the unusual terminology, and the necessary employment of a good deal of higher mathematics.

This is an affirmation that aeronautical engineering had come of age as a discipline in its own right – no longer was it a poor relative of its well-established mechanical, electrical, and civil counterparts.

In his own preface to the book, Thomson gives thanks to O'Gorman and Tizard, among others, but for some reason he does not acknowledge Bairstow, not even during his discussion of wind tunnels.[15] Thomson certainly knew Bairstow, and indeed the two had worked together considering future designs for commercial, civilian aircraft. One might wonder, therefore, if this omission was simply an aberration – or could it have been indicative of the strained relationship that was known to exist between those working at the Royal Aircraft Factory and those who worked at the NPL during the war? When Thomson's book was published in 1919, the *Reports and Memoranda (R&M) of the Advisory Committee for Aeronautics* were publicly available only up to those for 1915. These highly technical documents were all eventually assigned a report number but only after being passed as suitable for publication and distribution for general consumption.[16]

Thomson's book was commissioned, in effect, by Henry Tizard (then a full colonel at the Air Ministry), and fully supported during its writing by O'Gorman (a half-colonel) and many of Thomson's former associates at the Royal Aircraft Factory and the NPL. Not many could boast a father capable of reading the proofs of such a technical book, describing a new field of engineering, let alone one who could offer useful and constructive advice, but J.J. Thomson was hardly an average father when it came to understanding technical mathematics and new principles in engineering and design!

Thomson's treatise helps summarize what was known in aerodynamics and aeronautics, in a mathematical sense, in 1920, and it also gives some insight into the modus operandi of aeronautical R&D. Aircraft design teams comprised individuals with different skill sets, as O'Gorman relates (Thomson 1919, v–vi):

[15] Bairstow reciprocates in his publication by failing to mention Thomson.

[16] Prior to being assigned a report number, the reports held a 'T' designator at the Royal Aircraft Factory: T1234, for example.

> In a properly-balanced and well-ordered aircraft design office there are available ... specialists on the aerodynamic questions, engineers concerned with the preparation of stress diagrams, whose work is the basis on which to each part is ascribed its strength in proportion to the duty to be done, designers who have specialized on mechanical arrangements, and specialists on airscrew design tests and experiments.

He goes on to say that:

> In the drawing office proper concrete form is given to the thoughts, calculations, and conclusions of these men.[17] Moreover, as the best way to be sure is to have tried and verified, there is also, under the same control as the design branch, an engineering laboratory in which tests of every proposal, whether for the struts, links, ties, spars, or bracings, as well as tests on special methods of fixation between wood and wood, or between wood and metal parts and the like, are conducted.

O'Gorman goes on to describe the use of wind tunnels, particularly to assess lift and stability, and the construction of one or two prototype aircraft that could be used first for full-scale testing and then for destructive testing on all the major aircraft components. Only when all these tests were satisfactorily completed would the expense of setting up bespoke jigs and manufacturing lines be authorized.

By 1919, then, the process of how to go about successfully designing, testing, and mass-producing a new aircraft had matured. It was no longer individuals trying to play all the roles and cobble together something that might fly; it was an integrated process able to draw on the expertise of a number of individuals, each possessing a specific skill set, who were able to tap into the services provided by the various test facilities as required.

But what were the key questions in aeronautical engineering at this point in its evolution? Without doubt, mystery still surrounded the precise mechanism of aerodynamic lift and, indeed, all attempts to define mathematically the precise forces acting on an aircraft during the various modes of flight had drawn a blank. As Thomson states (Thomson 1919, 18):

[17]We do not know whether Thomson and O'Gorman were unaware of the contribution of women in this field, were aware of it but chose to ignore it, or were simply conforming to stereotypical phraseology.

> In spite of the enormous amount of work which has been done in aerodynamics and the allied science of hydrodynamics there is no mathematical theory by which the forces on even the simplest bodies can be calculated with accuracy. In consequence, aerodynamics has tended to become a mere collection of measurements of forces on bodies of various shapes without underlying principles.

The problem with known hydrodynamic theory was that much of it only related to incompressible, inviscid fluids, so it was not transferable to describe an aircraft in flight, where the compressibility and viscosity of the air are intrinsic properties that simply cannot be ignored. Work had been done prior to the war using airships to find the viscous drag on a body in flight, but nothing came out of these tests in relation to lift. Pressure-measuring techniques and devices were reasonably advanced, which was at least making data-gathering easier and more reliable. Well-made pitot/static tubes were able to measure velocity accurately, and the use of manometers connected to surface holes could provide useful data on pressure distributions over surfaces in an airflow, as we saw with Taylor's work. Scale effect was understood to some extent, making wind tunnel data more applicable and transferable to full-scale machines, although nearly all tests on models omitted the propeller, which inevitably led to significant inaccuracies.

In summary, aeronautics in Britain (and indeed elsewhere) in 1920 was predicated more on physical experiences, and data collected during them, than on what any theory was able to provide.

The realm of stress calculations is a good illustration of this limitation. How to calculate the loads experienced by engineered structures was a very mature field, so the related mathematics was well understood: given a box girder structure, for example, methods existed that enabled calculation of the stresses experienced by any part of it. Even if the girders of that structure had been made from aerofoil sections, the stresses could still have been calculated in a static situation with no wind present. Uncertainties only arose if an airflow was introduced, because now the exact position and magnitude of the acting forces became undefinable in a theoretical sense. Even experimentation and data capture could not cater for all the possible permutations of flight conditions, so the extremes of a safe flight envelope for an aircraft were difficult to define. Add to this the unknown aeroelastic forces, resonance effects, and cumulative fatigue issues, and it is easy to understand why aircraft were still susceptible to structural failure, even after the war.

That is not to say that a huge canon of related mathematics had not been accumulated: it had, with much of it sitting in the technical reports of the ACA and glimpses of it being seen in the contemporary general literature. The mathematics of aeronautics had certainly taken a quantum leap forward by 1920 compared with its state two decades before, and the stage was set for its natural evolution to cater for the aerial mass transportation of people and goods.

9

Conclusion

It is perhaps difficult to define the point at which the story of fixed-wing, powered flight began. The creation of these revolutionary aircraft was not dependent on one idea or one particular piece of technology, but rather required the conflation of numerous, disparate elements of engineering, mathematics, and materials science. It was not so much a revolution, then, as it was a convolution: a maturing and bringing together of existing know-how. Reliable gliders had been around for decades before the Wrights took to the air, so the aerofoil was not a new concept. Internal combustion engines had been available for nearly half a century. Light metals such as aluminium could be sourced and fashioned, and the crafting of wood and fabric were ancient arts. The design and construction of a viable craft therefore really came down to two constraints: could the necessary power-to-weight ratio be achieved alongside sufficient structural integrity, and could any conforming contraption be controlled once in the air?

One might expect that mathematics was an essential ingredient in solving these final aeronautical conundrums, but in many ways it was just the poor sibling of desire, certainly in those early days. The practical men just wanted to fly, and an unsolved quartic equation was never going to stop them, however much the likes of George Bryan protested. This accusation could even justifiably be levelled at the mathematicians who eventually took to the air: they, too, just wanted to fly. Yes, technically they had an excuse to be up there, but the bottom line was that it was as exciting as it was necessary.

The period in which all this was taking place is, therefore, significant. It was a pioneering and enthusing time in aviation. Few people had been given the opportunity to fly, and working in an aeronautical environment where cutting-edge advancements were being made almost daily must have heightened the desire to participate. That said, there were those for whom the thought of being taken skyward was unappealing: Hermann Glauert, for example, was a happy ground-dweller. For most of these academics, however, gaining one's wings was appealing, so to them any associated risk was a calculated one and was deemed worth taking. This is not conjecture: all who ventured skyward made their complicity in the act quite explicit. There was also an element of duty in all of this. The mathematicians were only at Farnborough at all because of a desperate desire to contribute to the war effort. One gets the impression that they might have felt it incongruous to avoid confronting some measure of danger, particularly since some of their number, such as Renwick and the Hill brothers, could empathize only too well with the perils of the battlefield, having spent some time at the front.

Duty and intrinsic curiosity aside, however, it is legitimate to ask whether the nature of these early test flights really warranted specialist participation by mathematicians and scientists. Could all flight tests actually have been delegated to army test pilots and observers? The answer is not entirely clear-cut. It is certainly the case that some experiments benefited from the fact that those conducting them had a measure of understanding of the academic technicalities and principles involved: Frederick Lindemann and his spinning trials, for example. But others were simply exercises in writing down figures from instruments: something that could have been done perfectly adequately by anyone who was literate and numerate. It seems reasonable to assert, therefore, that there was partial rather than unequivocal justification for allowing these academics to fly.

But what of the cost? We have seen that the lives of Ted Busk, David Pinsent, Keith Lucas, Hugh Renwick, and Bertram Hopkinson were all lost prematurely. Hopkinson was a hugely influential and important figure in academic engineering. Lucas, too, excelled in his own field of physiology. Pinsent's philosophical ties combined with his exceptional mathematical talent gave him great potential, as was true of Renwick. And it would be hard to overstate the impact that Busk might have had on British aviation had he not succumbed to the perils of flight. There was thus considerable cost to society beyond the personal human tragedy.

One is left to imagine what contributions these people would have made to mathematics, aeronautics, science, and society in general had they stayed alive till war was over. An entirely utilitarian and dispassionate view has to be that their sacrifices were an inevitable consequence of a period in which huge advances in practical and theoretical aeronautics were made in Britain, but one can never totally divorce the personal aspects of their stories from this analysis. They demonstrated extreme bravery, and the achievements of these men – the first to 'fly and apply' in fixed-wing, powered aircraft; true pioneers of aviation – were extraordinary.

It is perhaps worth considering from the perspective of a pilot what exactly justifies such accolades. Every modern-day test pilot always has a contingency plan in the event of catastrophic failure in flight – a plan that gives him or her a chance to survive. This might be an ejection seat or parachute, or it could be some measure of redundancy in critical systems. Pilots flying modern aircraft have an array of accurate, real-time telemetry to apprise them of position, orientation, speed, altitude, fuel state, engine performance, system malfunctions, load factor, and so on. They have stall, terrain, excessive load, and over-speed warnings, and the assistance of computers to help with every aspect of the task of operating an aircraft. Communication is not an issue. Weather can be monitored, predicted, and, if necessary, avoided.

Contrast all this with what faced the likes of Busk. Pilots back then had airframes that might fail structurally at any moment, only a handful of rudimentary instruments that were compromised by vibration or design issues, and unreliable power units. Open cockpits also exposed aircrew to the elements, with excessive noise and vibration being very distracting and posing significant discomfort. Even rudimentary operating techniques such as spin recovery were not properly defined for these early aviators. They were, indeed, a breed apart.

It may have been the changing role of aircraft that was the catalyst in bringing the importance of the mathematics into clearer focus. Flying straight and level between London and Manchester for pleasure put very different demands on an aircraft structure from, say, dog-fighting over the Somme. The sheer number of fatalities due to issues of control or structural failure made it imperative that more accurate, quantitative analysis became a mandatory and better-understood part of the aircraft design process.

During the first two decades of the twentieth century, there was an uncomfortable measure of uncertainty and debate around the actual processes that caused an aerofoil to produce its lift; David Bloor's book title sums it up nicely: it really was 'the enigma of the aerofoil'. Other factors that determined aircraft performance and integrity were also only partially understood: exactly how the position of the centre of pressure on a wing varied with phase of flight, for example, and the nature and magnitude of aeroelastic forces that could be encountered. These issues forced mathematicians and engineers to draw on known mathematics in established fields and consider what modifications were necessary to describe the aircraft in flight. Additionally, mathematics was certainly needed to inform the execution of certain manoeuvres or events: stalls, spins, and aerobatics being typical examples. Load factors and control were always at the forefront of these analyses.

The role of the literature in the story up to 1900 was modest. There was certainly interest among those intimately involved in aeronautics, but the reporting and exchange of mathematical developments was confined to *The Aeronautical Journal* and a smattering of articles in the more technical newspapers and scientific journals, such as *Engineering* and *Nature*. The turn of the century, however, brought with it a proliferation of aeronautics-related mathematical literature that would keep pace with the developments in hardware and flying techniques.

The expositions in question were diverse in both nature and form, and they were multifunctional. We are used to encountering literature that simply informs – passes on knowledge to the reader – so the various books and general newspaper articles that began to appear were to be expected. What was perhaps surprising, however, was the way in which mass-circulation publications such as *Flight* were being used as a medium for conducting open forum debates. Aeronautics in its new context was such a rapidly developing field that few topics within it had matured, and it was therefore not uncommon to find members of the general public making useful and significant contributions to progress. It is true that such people often had a mathematical pedigree (Joseph Hume-Rothery, for example), but for such people to be debating with, questioning, and suggesting advancements to those intimately involved in aeronautics is a fascinating aspect of this tale.

The literature clearly performed a number of additional functions. There was an element of lobbying in some articles, for instance: with allusions to national defence and world status, some journals provided a soap box from which political points could be declaimed. None exploited this more than the likes of Charles Grey and Noel Pemberton Billing, who, in their positions as editors of *The Aeroplane* and *Aerocraft*, respectively, were well placed to forward personal agendas or causes. Others were able to disseminate information and opinions more surreptitiously by using the cover of pen-names or influencing second parties to write what was necessary, as we perhaps see with Mervyn O'Gorman's enigmatic influence on aeronautical contributions carried in the *Times Engineering Supplement.*

The technical and operational data that was forthcoming, particularly in the pre- and post-war eras, was remarkable. Some articles were detailed enough to allow a reasonable attempt at the complete construction of an airframe. This was an age when the marketplace was awash with talented technical drawers who had honed their skills in rail, shipping, and other heavy industries, and they were ready to transition into the world of aviation. Things changed during the war itself, of course, when advances in design, in both airframes and engines, and developments in instrumentation and operating procedures had to be vetted to avoid potential use by the enemy.

How mathematics in the explicit sense integrated with all of this is interesting. While some of it would have been accessible, in an academic sense, to a general audience, much of it was clearly beyond the technical comprehension of all but a few. The latter implies that the literature was being used as a way for mathematicians and engineers to communicate with each other rather than as a tool for explaining aeronautical principles to the layperson – calculus and trigonometry do not sell newspapers. Offering up the detailed mathematics did, though, allow those reading it who were intimately involved in aircraft R&D to follow arguments, spot flaws, or identify potential developments while gaining better intuition than might otherwise have been the case. There was very little mathematics that was inherently new on show, but plenty of new applications and nuances were being uncovered and discussed.

The literature also hints at the close-knit and exclusive nature of the aeronautical community. The same names crop up time and time again as authors, editors, reviewers, board members, sponsors, operators, designers, manufacturers, trainers, educators, and trainees – all generally members of what might be called the 'aeronautical set'.

Among the pages of journals and reports, two generic strands of influence can be seen to emerge from the production lines of academia and practical engineering. Cambridge University was churning out a constant stream of intellectuals, augmented by other centres of academic excellence in London, Manchester, and beyond, while the technical colleges and schools, such as Finsbury, Crystal Palace, and those in the dockyards, provided a complementary stream: those who knew a slide rule from a straight edge. In many respects, then, the war was responsible for throwing together these two groups, with their disparate backgrounds and training, and forcing them to meld their skills to become a hybrid beast: the aeronautical engineer.

Some, though, refused to compromise. Bryan was never going to exchange his pen for a spanner. Equally, de Havilland would have found it difficult to sit in an office crunching numbers all day while an aircraft sat outside ready to be flown. But between the extremes was a group of individuals who resembled some mythological band: not so much Argonauts as Aeronauts, their quest being not to find a fleece but to uncover the mysteries of flight. Their Jason was Busk, who represented the pinnacle of what combining academic engineering talent with flying ability could offer in promoting aeronautical progress. The men who followed him were eclectic and no less effective. Stimulated by the pressing need to provide the RFC with the best possible aerial machines, they theorized, constructed, tested, and constantly re-evaluated. It was practical engineering overlaid on a foundation of mathematics and then tested for efficacy and safety.

Amid all of this, it is important to note that engineering mathematics was not a field abundantly populated by women in the early 1900s. There was little societal precedent, encouragement, or support for things to be otherwise. The women discussed in this narrative, however – Hilda Hudson, Letitia Chitty, and Beatrice Cave-Browne-Cave – certainly stand out as clear examples of those who swam against the prevailing tide of prejudice and conventional expectation.

They all benefited from familial stimuli and resources, which were almost obligatory prerequisites for any young woman who aspired to go up to Newnham or Girton, but infringing on the male-dominated world of mathematics also demanded other intrinsic qualities. These women were eminently capable of original and independent mathematical thought, and they had the strength of character to overcome the tradition that stood in their way. We need only look to Hudson and her treatise on Cremona transformations or to Chitty and her later eminence and expertise in the field of stresses in arches and dams to find proof of the natural mathematical talent these women possessed.

Their mathematical flexibility was also testament to their broad understanding of the subject. Hudson, a geometer at heart, tackled applied mathematics to help the war effort. Cave-Browne-Cave engaged in statistical analysis with Karl Pearson and transitioned to stress analysis with Sutton Pippard. Chitty made the switch from mathematics to the mechanical sciences at Cambridge and became a respected civil engineer. They were also groundbreakers: Hudson was the first woman to deliver a paper at an ICM, Chitty was the first woman to be placed in the first class in the Mechanical Sciences Tripos, and Cave-Browne-Cave was one of the few women to sole-author an ACA technical report.

Objectively, much of the mathematics these women were doing while working for the Admiralty during the war could not be described as exceptional, but it was certainly necessary. With each new design of aircraft came a demand for the calculations that would determine its structural integrity and limitations. Without competent and dedicated mathematicians, the new breed of industrial aeronautical engineers would have had to rely entirely on the rather blunt tools of judgement and experience. That said, while the bulk of the mathematics was well established, its specific application to aircraft was not. Here lay the challenge.

For female mathematicians in those early days of fixed-wing aeronautics, the pathway leading to a career in applied mathematics outside of education was, as we have seen, strewn with all manner of obstacles, but Hudson, Chitty, and Cave-Browne-Cave negotiated them admirably. Did the demands of the war, and the urgency those demands brought with them, assist their career progressions? The conflict certainly created an employment vacuum as men were conscripted in ever-increasing numbers, but there is no doubt that these women seized their opportunity to infiltrate the male bastion of engineering. Their calculations, expertise, and diligence undoubtedly prevented many unsafe designs moving from the drawing board into production. This was their practical legacy. But these women represented more than just technical checks and balances in the chain of aircraft design and production: they were pioneers who demonstrated that it was possible for women to overcome the dogmatic, institutionalized prejudices of the time. They earned the right to stand tall as credible applied mathematicians during the genesis of aircraft stress analysis.

FINAL THOUGHTS

The British public, particularly those involved in the various heavy industries, had been used to seeing mathematics in the literature throughout the nineteenth century. Mechanical and civil engineering, in particular, were very well established, and publications such as *The Engineer* had detailed the mathematics of construction, shipping, the railways, and so forth for their readerships for decades. There were no dedicated aeronautical journals since aeronautics was not in itself considered an engineering discipline. After all, fixed-wing, powered flight was a thing of fantasy at the time. Yes, there were balloons and gliders, and there were general articles covering events, theory, and developments relating to these vehicles, but any mathematics that was seen tended to be more in the linked fields of fluid dynamics and stress analysis.

All that began to change, though, in the first decade of the twentieth century, as aeronautics began to establish itself as its own engineering discipline. The success of the Wright brothers in 1903 proved that fixed-wing, powered flight was possible, and the race to build faster, stronger, longer-range, and more easily controlled machines began. This competition encouraged the fusion of two skill sets: the practical of the industrial engineers, who were used to getting their hands dirty; and the theoretical of the thoroughbred academics. The flying mathematicians, engineers, and scientists that made up the Chudleigh lot were the product of this amalgamation.

This period saw an evolution in the mathematics needed to describe and predict the behaviour of these new flying machines, and all of this was of course expounded on in the literature. For those more intimately involved with, or interested in, the theory, seminal books were published by the likes of Lanchester, Bryan, and Berriman. For those allied to the exciting new flying clubs and associations, journals and magazines covered the developments, often going into great technical detail. Mathematical debate was conducted under the public's gaze in these publications, with academics, industrialists, military men, and interested amateurs all playing a role.

The literature really blossomed in 1909. The technical editors of new publications such as *Flight* were instrumental in bringing the mathematics associated with aviation to the masses. It is notable that during this period nobody was quite sure what made the aerofoil produce lift. Stability was a challenge, as were power-to-weight issues due to the relatively modest performance of aero engines. Aerodynamic stresses were not properly understood. The war undoubtedly provided the driving force to advance these things apace, however, and journals and magazines, combined with the ACA's technical reports, kept the British public up to date with mathematical advancements as aeronautical engineering established itself as an important area of R&D. Its roots were firmly anchored in applied mathematics, and it provided the foundations upon which today's mature field of aeronautics was built.

Despite all the mathematics, however, one gets the impression that those first two decades of fixed-wing aeronautics in Britain were predicated on a framework of practical engineering coupled with a pioneering spirit of adventure and experimentation. Nevil Shute perhaps hints at this through the words of one of the characters in chapter 3 of *Stephen Morris*, a short story set just after World War I: 'Differential equations won't help you much in the design of aeroplanes – not yet, anyhow' (Shute 1961). Shute was himself an aeronautical engineer, and he was clearly fully aware that there was still much to learn about the mathematics that would eventually describe the finer aspects of flight. But we should not take his assertion as a reason to dismiss the importance of the contributions of the flying mathematicians and their ground-based counterparts. Their endeavours laid the foundations for stability and control, for structural integrity and design, and for standard operating procedures. Their experiments and calculations allowed the construction of the first and second generations of British aircraft, with the latter being far more capable and robust than the former in every respect. Mastering powered flight was an exciting new pursuit and it caught the imagination of the general public, who were fed a constant stream of news and information via the contemporary literature. The development of the mathematics of aeronautics was part of that story, and while the complex details were often hidden away in technical reports, the broad concepts were on display for public delectation, debate, and enlightenment.

During the third decade of the twentieth century, Geoffrey Hill set about designing an aircraft with a more forgiving wing: one that might protect pilots from the perils of the stall and spin. Because of its appearance, he named it the 'pterodactyl'. Walking through the gates of the Royal Aircraft Establishment now, one might notice the very same pterosaur atop the crown of authority of the coat of arms on display. A more likely reason for its presence here than its shape is the fact that it represents the earliest vertebrate known to have evolved powered flight.

Just as fitting is the motto chosen to accompany the emblem: 'Alis Apta Scientia' – 'Wings Through Knowledge'. This is a sentiment that resonates with the very ethos of the Factory. It really was the place where essential knowledge was garnered and employed by some of Britain's finest mathematicians to allow the viable construction of the country's first tranche of powered flying machines. Those working there created a deep well of information on which the authors of the contemporary literature could draw to tell the masses the tale of the birth of British aeronautical engineering. Through their pursuits they also defined the very essence of what made flying both exciting and dangerous during those pioneering years of aviation. Just as the Royal Aircraft Establishment lives on, so too should our memories of the incredible exploits, contributions, and achievements of the talented and brave flying mathematicians of World War I, to whom this book is respectfully dedicated.

Appendix

Engineering and Aerodynamic Issues

THE FUNDAMENTAL CHALLENGES IN 1900

The primary forces acting on an aircraft in flight are thrust, drag, weight, and lift. Combined with any external stimuli, such as atmospheric conditions, how these forces interact dictates how an aircraft moves and behaves in flight at any given instant. Of the four parameters, it was the thrust in relation to an aircraft's weight that was the annoying limitation in aeronautics at the turn of the twentieth century.

POWER-TO-WEIGHT RATIO

Solving the thrust versus drag issue that faced early aeronautical engineers came down to the design and manufacture of an engine and propeller combination that could produce sufficient power (thrust) to overcome the forward resistance to motion of the airframe (drag), while being light enough – when combined with its fuel supply and the airframe itself (weight) – to be raised into the air by the vertical component of the force generated by the wings (lift). What one might think of as conventional engineering R&D was therefore all that was needed to inform the production of more efficient power units that were able to increase the power-to-weight ratio of any proposed airframe/engine combination. In tandem with the evolution of lighter airframes that could easily slip through the air while still being robust enough not to fall apart under any applied loads, advances in aero engine technology would be the key that finally unlocked the revolution in aeronautics.

Unlike thrust, drag was a poorly understood concept despite all the early theoretical work that had been done in the field of fluid dynamics and the practical experimentation conducted in the first wind tunnels. The Aeronautical Society of Great Britain, funded by donations and subscriptions, had its first official gathering on 27 June 1866, and one of its founding members, Francis Wenham, went on to design and operate the first wind tunnel in 1871. As early as 1858, he had performed tests on a multiplane glider, concluding that a cambered wing generated most of its lift from the front section: a fact that underpinned the early preference for wings with a high aspect ratio.[1]

The hope in 1900, therefore, was that engine technology would move on apace and negate any immediate necessity to grapple with the nuances of the drag issue. It is remarkable, and certainly worth noting here, that many scientists and engineers throughout the 1800s actually believed that powered flight would always be an impossibility. One imagines that their reluctance to even contemplate success in such a venture was based on their ignorance regarding the full potential of the aerofoil combined with the unimpressive progress being made in the field of propulsion: a steam design was clearly never going to be fit for purpose, although Hiram Maxim did his best to disprove this assertion!

Once engines became lighter, more reliable, and capable of generating increasing amounts of power with each iteration, the pressing need to discover and define the causes of aircraft drag diminished, of course. Figure A.1 shows the engine that finally provided the required power for its weight. The four-cylinder, in-line design used an aluminium crankcase but had no carburettor, spark plugs, or fuel pump. It produced 12 horsepower and weighed 180 lbs.

STRUCTURAL INTEGRITY

The physical construction of an airframe was a well-established art by 1900: various manifestations of gliders had been flying for decades. What was not fully understood, however, was how and where all the aerodynamic forces acted; nor was it always possible to calculate their magnitude. Simply 'over engineering' the structure to make it physically robust was not a solution because such a strategy inevitably resulted in frames that were completely impractical due to their weight. Whirling arms and wind tunnels were employed during the late 1800s in an attempt to gather data using

[1]Following royal approval from King George V in 1918, the Aeronautical Society was renamed the Royal Aeronautical Society.

Figure A.1 The Wright brothers' engine.

scale models, and while they certainly provided some insight, they did not explain the whole picture.

The basic procedure that was adopted was to set a defined limit on the strength required and then carry out the necessary calculations to ensure that each part of an airframe could withstand stresses up to that limit. The whole process was centred around known structural mechanics. Some of the principles had to be slightly modified for specific application to aircraft, but the basic mathematics and techniques were in place well before aircraft were being designed and flown. The problems in stress analysis therefore came down to the vague understanding of the actual forces being encountered in flight. Where did the force vectors act? How large could they become? What sort of specific stresses could an aircraft encounter that static structures would not suffer?

In these early days, then, there was a constant dialogue between those who were performing theoretical calculations and those who were carrying out actual air tests. If there was one phenomenon that was very poorly recognized and understood, it was the aerodynamic effect that would become known as 'aeroelasticity'.

AEROELASTICITY

The way airflow interacts with an aircraft's flexible frame can produce some interesting modes of motion within the structure itself, and these forced movements or oscillations fall under the general heading of aeroelastic effects. They can be thought of as the result of the interaction between inertial, structural, and aerodynamic forces at any given moment for an aircraft in flight. The phenomena were known by observation to the early aeronautical engineers but were not fully understood. In modern aircraft design, computer simulations can be used to determine the nature of these effects and either inform design modifications or define performance limitations to keep the aircraft within a safe envelope of operation. During the early years of aviation, however, design and certification methods were rather more crude: aircraft were often simply flown at their maximum speeds to provide some measure of reassurance regarding their aeroelastic resilience.

'Flutter' is one of the most common aeroelastic phenomena that aircraft encounter, often occurring in the main wings. Wings in general have two degrees of freedom: they can bend span-wise, and they can rotate in pitch. There are, therefore, rotational and vertical motions combining. Issues arise when the frequencies of these two motions coalesce in flight to create a resonance that can grow in amplitude until the structural strength is no longer sufficient to hold the wing together. This potential resonance is not confined to the main wings: it can happen in the tail assembly, in control surfaces such as ailerons and rudders, and in propeller blades. The first official record of a flutter incident was in 1916, when the torsion in the fuselage of a Handley Page O/400 bomber coupled with an antisymmetric elevator movement to cause significant handling problems for the pilot.

As the maximum speeds of aircraft began to increase, so too did the number of control surface incidents involving flutter; wing-aileron flutter was arguably the most common.[2] As discussed on p. 147, the British mathematicians Leonard Bairstow, Arthur Fage, and Robert Frazer, all working at the NPL in Teddington, were at the forefront of trying to devise solutions to these problems, in conjunction with the test pilots at the Royal Aircraft Factory. Indeed, one of the most important structural modifications that Frazer

[2]As described by aerodynamicist Arthur Collar (Collar 1978a).

Figure A.2 Handley Page O/400 bomber.

devised was the use of mass balances on ailerons to help prevent wing flutter (Pugsley 1961).

It would be the shift towards monoplane wing designs and higher air speeds after World War I, however, that prompted an increase in the number of main-wing flutter issues. The man who most advanced understanding of the subject in the immediate post-war era in Britain was undoubtedly Robert Frazer, and he was assisted in this pursuit by Cambridge graduate Arthur Collar at the NPL at the end of the 1920s. Collar would publish 23 papers in the R&M series, and he laid the foundations of advanced R&D in the field of aeroelaticity in Britain. The full extent of his contribution can be appreciated by inspection of the bibliography in mechanical engineer Dr R.E.D. Bishop's comprehensive *Biographical Memoir* of Collar (Bishop 1987).

From a pilot's perspective, any excessive airframe vibration in flight can be most disconcerting. In the modern era it is generally caused by some sort of imbalance in the engine(s) that is transferred to the airframe, or by something such as an external panel becoming loose. If one gets into an over-speed situation, this may also induce some shaking. In the early days of flight, the circumstances were very different. The onset of the twisting of the wings or control surfaces was invariably rapid and confusing. Before it was possible for the pilot to make any sort of diagnosis or corrective action, it was inevitably too late to prevent structural failure. More often than not, such events led to a crash and the death of the aircrew on board. This meant that feedback from aeroelastic experiences in the air was relatively scarce. Post-crash investigation techniques and procedures were not as revealing as they are now, so looking at what remained of an aircraft following a crash did not always tell the whole story of the sequence of events that caused a particular airframe to break up in flight. As was evident concerning the events that killed Pinsent and Renwick, the only real clues as to what happened came from eyewitnesses on the ground.

SCALE EFFECT

Studying effects such as flutter would have been a far safer, more cost-effective, and expeditious business had it been possible to conduct the bulk of aerodynamic tests using models in wind tunnels rather than full-size aircraft in the air. Unfortunately, however, the rather annoying, perplexing, and inconvenient issue of 'scale effect' stood in the way. The essence of this phenomenon is that when considering the fluid dynamical properties associated with an aircraft moving through the air, certain aspects of the interaction between the fluid (the air) and the aircraft differ depending on the size of the craft. Marginal differences are of little consequence, but when comparing a model aircraft with a full-sized version, the discrepancies become substantial and significant.

The crux of the matter for an aerofoil in flight is the behaviour of the air in the boundary layer: that is, the air in the immediate proximity of the aerofoil surface. In this region there is a rapid change in the air's velocity depending on its distance from the surface. The relative movement between adjacent thin sheets of air moving at different velocities is what determines any propensity for turbulence to be created in the boundary layer. The viscosity of the air also significantly influences the dynamics in this region.

The first person to put all these concepts together in a coherent way was the Anglo-Irish physicist and mathematician George Stokes in 1851; his work was clearly not inspired by the field of aeronautics, which was yet to be established. Osborne Reynolds, 7th Wrangler in the Mathematical Tripos of 1867, would be the man to exploit the theory more fully. A professor of engineering at Owens College in Manchester from 1868 until the end of his academic career, Reynolds published an important paper on the subject in 1883 (Reynolds 1883). Such was the respect for his work on this aspect of fluid dynamics that its key parameter – the ratio of inertial forces to viscous forces within a fluid – was named after him. The 'Reynolds number', *Re*, gives some intuition into the nature of fluid flow in the boundary layer. It is defined as follows:

$$Re = \frac{\rho u L}{\mu},$$

where ρ is the fluid's density, u is the fluid's velocity relative to an object, μ is the fluid's dynamic viscosity, and L is some defined linear dimension of the object. When *Re* is low, viscous forces dominate proceedings. This promotes a laminar (smooth) flow. At high values of *Re* the opposite is true, with the large inertial forces causing turbulence and instability in the fluid.

For an aerofoil, the linear dimension used in the equation is often the chord length of the wing. Imagine if one was to calculate *Re* for a real aircraft and then for a scale model of that aircraft. All other things being equal, the values of *Re* obtained would differ by the factor of the scaling. This scale effect therefore had to be considered when attempting to study a model and relate its performance in the wind tunnel to that of the real aircraft in flight.

Confounding this comparison are the different densities of air in which the real aircraft might find itself while flying. One might think that compensating for the smaller L in the wind tunnel might be a simple case of increasing u by the same factor. But if, say, the aircraft was flying at 100 miles per hour and the scale factor was 10, this would mean that the wind speed in the tunnel would need to be 1,000 miles per hour to compensate. But that air speed is supersonic, which has all manner of implications that make this potential solution impractical. And when you combine this with the differences in the data gathered between real and model aircraft caused by the effect of the tunnel walls on vortex production for the latter, it is hardly surprising that the mathematicians and engineers attempting to make sense of it all had their work cut out. A comprehensive – and geographically more inclusive – discussion about the scale effect debate that raged throughout the World War I period and beyond can be found in Hashimoto's works mentioned in the preface: Hashimoto (1990) and Hashimoto (2000).

MATERIALS SCIENCE

One of the most important factors in determining the structural integrity of an aircraft is, of course, the material from which it is constructed. Most early aircraft structures tended to be made from various species of wood, predominantly spruce, and there were many reasons for this: it was relatively light and inexpensive, it was easy to source and fashion, and it could be made flexible or rigid. Metals were also an option in the early 1900s, of course. Aluminium alloys were particularly popular as a means of reducing weight and for components that required extra strength. Some of the most important constituents, particularly in the biplane designs, were the struts that held the wings both apart and together. Combined with the various rigging wires, it was these members that were often exposed to the greatest stresses, and it was therefore essential that they were robust enough to cope during any phase of flight.

The fabrics chosen to cover the wings and fuselage were also important: they had to be strong and durable. Much of the work done by the chemists at the Royal Aircraft Factory involved developing the most effective dopes to strengthen these flimsy fabrics and protect them from the elements.

Following the war, the power of engines increased rapidly, and there was an inevitable move towards monoplane wings and metal frames, with cloth coverings being replaced by metal panelling; a discussion of this evolution can be found in Jakab (1999). Again, I think the uncertainty surrounding the distribution and magnitude of loads on the aircraft structure during the various phases of flight had a serious impact on the ability of design engineers to anticipate the optimum strength required of components: making them as light as possible while still not liable to fail in flight. The old adage that 'a chain is only as strong as its weakest link' must have been in the back of every pilot's mind when he or she was pulling out of a dive in those early days of aviation.

BRYAN'S RESISTANCE DERIVATIVES

This book has discussed the contributions of mathematicians, engineers, and scientists in tackling the engineering and aerodynamic challenges surrounding the design and construction of the first generation of powered aircraft. It is perhaps useful, therefore, to offer some basic insight into the mathematics developed by Bryan and used by Busk during the latter's attempts to tackle the key issue of

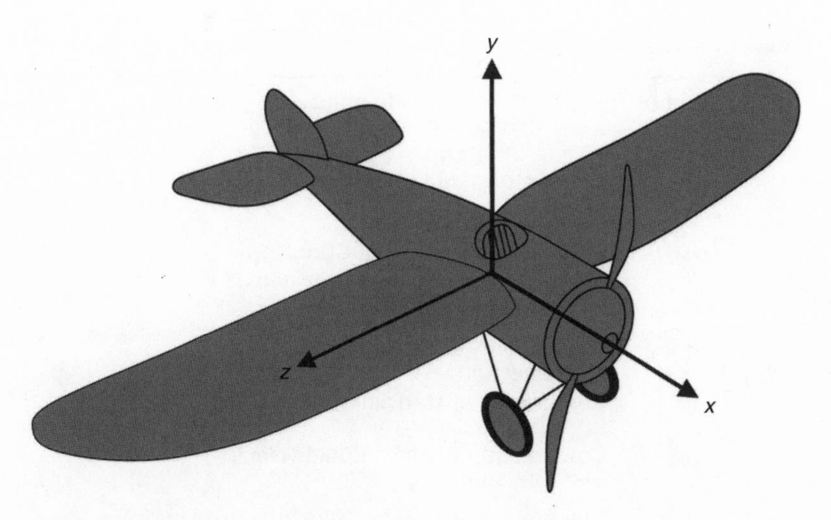

Figure A.3 Aircraft axes.

aircraft stability and control. This section thus attempts to encourage some intuitive understanding of Bryan's 'resistance derivatives', and how he arrived at his stability equations. The nomenclature and definitions that he used are shown in table A.1. This is by no means an attempt to reproduce the entire contents of the relevant chapters in Bryan's book, *Stability in Aviation* (Bryan 1911), but more an overview that offers some insight into his approach towards mathematically defining the design criteria that would result in an inherently stable aircraft.

Bryan's entire argument revolved around knowledge of the forces imparted on an aircraft by the components of air resistance in the direction of, and about, the three mutually orthogonal aircraft axes, as depicted in figure A.3. If we consider the longitudinal x-axis, for example, a typical value that needed to be defined was X. He asserts that X must be some linear function of the small quantities u, v, w, p, q, and r, and can therefore be expressed as such:

$$X = X_0 + uX_u + vX_v + wX_w + pX_p + qX_q + rX_r, \qquad \text{(Eqn I)}$$

where X_0 is the steady-state resistance component.

Since there are six of these small components in addition to X_0 – and six major resistances/couples, X, Y, Z, L, M, and N, which are, respectively, related to each of the former – the permutations generate 36 possible contributions: X_u, Y_w, Z_r, and so

Table A.1 Bryan's nomenclature and definitions.

Symbol	Meaning
W	Weight of aircraft
g	Acceleration due to gravity
H	Propeller thrust
A	Moment of inertia about the x-axis
B	Moment of inertia about the y-axis
C	Moment of inertia about the z-axis
u	Component of translational velocity in x-direction
v	Component of translational velocity in y-direction
w	Component of translational velocity in z-direction
p	Component of angular velocity in x-direction
q	Component of angular velocity in y-direction
r	Component of angular velocity in z-direction
X	Force of air resistance in x-direction
Y	Force of air resistance in y-direction
Z	Force of air resistance in z-direction
L	Rotational couple about the x-axis
M	Rotational couple about the y-axis
N	Rotational couple about the z-axis

forth. The symmetry of the system, however, allows this number to be reduced to 18 unique, potential influences; these were Bryan's 'resistance derivatives'.

Equations of motion can be defined in each orthogonal plane for both translational and rotational accelerations (Bryan 1911, 23). In the longitudinal axis, for example, applying Newton's familiar equation $F = Ma$ to an aircraft that has a downward pitch angle of θ to the horizontal and an engine and propeller that produce thrust H, the equation of motion is

$$W\sin(\theta) + H - X = W\left(\frac{\mathrm{d}u}{g\,\mathrm{d}t} + \frac{qw}{g} - \frac{rv}{g}\right). \qquad \text{(Eqn II)}$$

Here, on the left-hand side we have the component of weight acting in the x-direction, plus the component of thrust, minus the resistance of the air, giving the net acting force. This is equal to the mass (expressed as W/g) multiplied by the various accelerations relevant to motion in the x-direction. Bryan's challenge was then to find ways to simplify things. First he removes resistance components and couples that have zero effect in the direction of motion under consideration; in the longitudinal axis, for example, this gives a simplified form of (Eqn I) for the X components of resistance:

$$X = X_0 + uX_u + vX_v + rX_r.$$

He also considers the situation under steady-state conditions for an aircraft at a downward pitch angle of θ_0, producing a thrust H_0; this is a situation where there is no net force acting, so

$$W \sin(\theta_0) + H_0 - X_0 = 0.$$

Furthermore, he assumes that during small oscillations in pitch, the aircraft's pitch angle, θ, will differ from θ_0 by a small quantity, ε, thus

$$\sin(\theta) = \sin(\theta_0) + \varepsilon \cos(\theta_0).$$

These new equations, for a small change in thrust δH, can be used to simplify (Eqn II) to yield

$$W\left(\frac{\mathrm{d}u}{g\,\mathrm{d}t}\right) = W\varepsilon \cos(\theta_0) + \delta H - X_0 - uX_u - vX_v - rX_r. \quad \text{(Eqn III)}$$

Bryan repeats this process for all three aircraft axes for translations and rotations, which results in six equations: the first group of three define longitudinal oscillations in the xy-plane, and the second group of three define lateral oscillations. He then turns to the crux of his argument: how these equations of motion relate to the actual stability of an aircraft. In the case of longitudinal stability, he assumes that u, v, r, and ε are proportional to some exponential function, $\mathrm{e}^{\lambda t}$. In other words,

$$\frac{\mathrm{d}u}{\mathrm{d}t} = \lambda u, \qquad \frac{\mathrm{d}v}{\mathrm{d}t} = \lambda v, \qquad \frac{\mathrm{d}r}{\mathrm{d}t} = \lambda r, \qquad \frac{\mathrm{d}\varepsilon}{\mathrm{d}t} = \lambda \varepsilon.$$

Equation (Eqn III) now becomes

$$\left(\frac{W\lambda}{g} + X_u\right)u + X_v v + \left(-\frac{W}{\lambda}\cos(\theta_0) + X_r\right)r = \delta H.$$

If propeller thrust is now taken to be independent of aircraft velocity, which is a reasonable simplification, then the right-hand side of this equation becomes zero, giving

$$\left(\frac{W\lambda}{g} + X_u\right)u + X_v v + \left(-\frac{W}{\lambda}\cos(\theta_0) + X_r\right)r = 0.$$

This equation, combined with the two similar equations relating to longitudinal oscillations, allows Bryan to construct a 3×3 determinant with horizontal components that are the coefficients of u, v, and r in each equation, respectively. Evaluating this determinant produces a quartic in λ:

$$\boxed{\mathfrak{A}_0\lambda^4 + \mathfrak{B}_0\lambda^3 + \mathfrak{C}_0\lambda^2 + \mathfrak{D}_0\lambda + \mathfrak{E}_0 = 0}$$

We now therefore have some intuition regarding the place and relevance of the resistance derivatives within Bryan's analysis. In the quartic equation, $\mathfrak{B}_0$, for example, represents the quantity

$$g(CW(X_u + Y_v)) + W^2N_r.$$

So the quantities X_u, Y_v, and N_r are buried within the coefficients of the quartic, thereby influencing the nature of the roots of that equation.

This is important because one of Bryan's crucial assumptions was that any small disturbances would be proportional to $\mathrm{e}^{\lambda t}$. In other words, the disturbances were influenced by the sum of four exponential components, each with a root in its index:

$$\boxed{a_1\mathrm{e}^{\lambda_1 t} + a_2\mathrm{e}^{\lambda_2 t} + a_3\mathrm{e}^{\lambda_3 t} + a_4\mathrm{e}^{\lambda_4 t}}$$

Clearly, if any of the roots is real and positive, then disturbances will increase indefinitely with time, leading to instability. Alternatively, if the roots are all real and negative, disturbances will decrease with time, leading to stability of motion.

When considering the pair of complex roots $\alpha \pm \beta$, Bryan assumed disturbances to be of the form $\mathrm{e}^{\alpha t}(p\cos(\beta) + q\sin(\beta))$. If the real part of the root, α, is positive, there is instability, whereas a negative real part, $\alpha = -\gamma$, i.e. $\mathrm{e}^{-\gamma t}(p\cos(\beta) + q\sin(\beta))$, will result in a damped oscillation with what Bryan called a 'coefficient of subsidence', γ.

In summary, Busk had to find ways of measuring the resistance derivatives for any given aircraft in order that the coefficients of Bryan's quartic equation that related to that specific aircraft could be evaluated. With these numbers in place, the nature of the roots of that equation could be determined, and these predicted the stability or otherwise of that aircraft. Given time and experience, it is likely that Busk would have developed his own intuition regarding the potential impact of various design feature modifications on the equations, which would have made the process slightly less tortuous. It is easy to appreciate the potential value of being able to use wind tunnels in this pursuit, hence the pressing need at the time to understand scale effect.

Cast of Characters

Adrian, Edgar D. (1889–1977)
Aston, Francis W. (1877–1945)
Ayrton, William (1847–1908)
Babbage, Charles (1791–1871)
Baden-Powell, Baden (1860–1937)
Baden-Powell, Robert (1857–1941)
Bairstow, Leonard (1880–1963)
Berriman, Algernon E. (1883–1959)
Berry, Arthur (1862–1929)
Betz, Albert (1885–1968)
Birkbeck, George (1776–1841)
Blériot, Louis C.J. (1872–1936)
Bohr, Niels H.D. (1885–1962)
Boltzmann, Ludwig E. (1844–1906)
Booth, Harris (1884–1943)
Borchardt, Carl (1817–1880)
Brancker, William S. (1877–1930)
Brillouin, Marcel (1854–1948)
Brodetsky, Selig (1888–1954)
Browning, John (1831–1925)
Bryan, George H. (1864–1928)
Busk, Edward T. (1886–1914)
Busk, Hans (1894–1916)
Busk, Mary (1854–1935)
Byron, George G. (1788–1824)
Capper, John E. (1861–1955)
Castigliano, Alberto (1847–1884)
Cauchy, Augustin L. (1789–1857)
Cave-Browne-Cave, Beatrice (1874–1947)
Cave-Browne-Cave, Frances (1876–1965)
Cave-Browne-Cave, Thomas (1885–1969)
Cayley, Arthur (1821–1895)
Cayley, George (1773–1857)
Challis, James (1803–1882)
Chandler, Dorothy (18??–19??)
Chanute, Octave (1832–1910)
Chisholm, Grace (1868–1944)
Chitty, Herbert (1863–1949)
Chitty, Letitia (1897–1982)
Churchill, Winston L.S. (1874–1965)
Clapeyron, Émile (1799–1864)
Clark, Thomas W.K. (1873–1941)
Cockroft, John D. (1897–1967)
Cody, Samuel (1867–1913)
Collar, Arthur R. (1908–1986)
Crawford, Alexander (1790–1856)
Cremona, Luigi (1830–1903)
D'Alembert, Jean le Rond (1717–1783)
Darwin, Horace (1851–1928)
Davies, Thomas S. (1795–1851)
Day, Maurice C. (18??–1914)
De Havilland, Geoffrey (1882–1965)
D'Ocagne, Philibert M. (1862–1938)
Duchêne, Émile A. (1869–1946)
Dunne, John W. (1875–1949)
Eiffel, Gustave (1832–1923)
Euler, Leonhard (1707–1783)
Ewing, James A. (1855–1935)
Fage, Arthur (1890–1977)
Fairey, Richard (1887–1956)
Farren, William (1892–1970)
Fawcett, Philippa G. (1868–1948)
Ferber, Ferdinand (1862–1909)
Forrest, James (1825–1917)
Forsyth, Andrew R. (1858–1942)
Fowler, Henry (1870–1938)
Frazer, Robert A. (1891–1959)
Garner, Harry M. (1891–1977)
Glauert, Hermann (1892–1934)
Glazebrook, Richard T. (1854–1935)
Goodden, Frank W. (1889–1917)
Green, Frederick M. (18??–19??)
Greenhill, George (1847–1927)
Gregory, Olinthus G. (1774–1841)
Grey, Charles G. (1875–1953)
Haldane, Richard B. (1856–1928)
Hall, Arnold A. (1915–2000)
Handley Page, Frederick (1885–1962)
Hardy, Godfrey H. (1877–1947)
Hawker, Harry G. (1889–1921)

Helmholtz, Hermann von (1821–1894)
Henderson, David (1862–1921)
Herman, Robert A. (1861–1927)
Hermite, Charles (1822–1901)
Hill, Archibald V. (1886–1977)
Hill, Geoffrey T.R. (1895–1955)
Hill, Micaiah J.M. (1856–1929)
Hill, Roderic M. (1894–1954)
Hopkinson, Bertram (1874–1918)
Horsley, Oswald (1893–1918)
Howard, H.B. (18??–19??)
Hudson, Hilda P. (1881–1965)
Hudson, Ronald W.H.T. (1876–1904)
Hume-Rothery, Joseph H. (1866–19??)
Hutchison, Mary (18??–19??)
Hutton, Charles (1737–1823)
Jones, Bennett M. (1887–1975)
Jones, Reginald V. (1911–1997)
Keith-Lucas, David (1911–1997)
Kennedy, Alexander B.W. (1847–1928)
Lamb, Horace (1849–1934)
Lanchester, Frederick W. (1868–1946)
Landau, Edmund (1877–1938)
Lang, Eleanor (18??–19??)
Langley, Samuel P. (1834–1906)
Ledeboer, John H. (1883–1930)
Lilienthal, Otto (1848–1896)
Lindemann, Frederick (1886–1957)
Lovelace, Ada (1815–1852)
Lucas, Keith (1879–1916)
Lutyens, Edwin L. (1869–1944)
Lutyens, Lionel F.D. (1894–1918)
Macaulay, William H. (1853–1936)
Maudslay, Henry (1771–1831)
Maxim, Hiram (1840–1916)
Maxwell, James C. (1831–1879)
Mayo, Robert H. (1891–1957)
McKinnon Wood, Ronald (1892–1967)
Meitner, Lise (1878–1968)
Menabrea, Luigi F. (1809–1896)
Navier, Claude-Louis (1785–1836)
Ogilvie, Alec (1882–1962)
O'Gorman, Mervyn J.P. (1871–1958)
Parke, William (1889–1912)
Pearson, Karl (1857–1936)
Pemberton Billing, Noel (1881–1948)
Perry, John (1850–1920)
Pilcher, Percy S (1866–1899)
Pinsent, David H. (1891–1918)
Pippard, Sutton (1891–1969)
Pisati, Laura (18??–1908)
Powell, Baden (1796–1860)
Prandtl, Ludwig (1875–1953)
Pritchard, Laurence (1885–1968)
Rankine, William J.M. (1820–1872)
Reissner, Hans (1874–1967)
Relf, Ernest (1888–1970)
Renwick, Hugh A. (1891–1918)
Reynolds, Osborne (1842–1912)
Roe, Alliott V. (1877–1958)
Routh, Edward (1831–1907)
Rowe, Richard C. (1853–1884)
Schottky, Friedrich H. (1851–1935)
Schwarz, Karl H.A. (1843–1921)
Scott, Charlotte A. (1858–1931)
Segre, Corrado (1863–1924)
Short, Frank (18??–19??)
Shur, Issai (1875–1941)
Siemens, Werner von (1816–1892)
Sopwith, Thomas O.M. (1888–1989)
Soreau, Rodolphe (1865–1936)
Southwell, Richard V. (1888–1970)
Spooner, Stanley (1856–1940)
Squier, George O. (1865–1934)
Stanton, Thomas (1865–1931)
Stephens, Arthur V. (1908–1992)
Stevens, H.L. (18??–19??)
Stokes, George G. (1819–1903)
Strutt, John W. (1842–1919)
Sueter, Murray F. (1872–1960)
Taylor, Geoffrey I. (1886–1975)
Templer, James L.B. (1846–1924)
Thompson, Silvanus P. (1851–1916)
Thomson, George P. (1892–1975)
Thomson, Joseph J. (1856–1940)
Thurston, Albert (1891–1967)
Tizard, Henry T. (1885–1959)
Usborne, Neville (1888–1916)
Walker, George (1734–1807)
Webb, Robert R. (1850–1936)
Weierstrass, Karl T.W. (1815–1897)
Wenham, Francis H. (1824–1908)
Wilde, Henry (1833–1919)
Williams, William E. (1881–1962)
Wilson, Walter G (1874–1957)
Winbolt, John S. (1841–1903)
Wittgenstein, Ludwig (1889–1951)
Wright, Orville (1871–1948)
Wright, Wilbur (1867–1912)
Young, William H. (1863–1942)

Glossary and Aircraft Schematic

Aeroelasticity. As an aircraft moves through the air it experiences aerodynamic forces that may interact with its inertial and elastic properties. These interactions can lead to serious problems that may cause parts of the aircraft structure to warp, twist, or flutter. *Aeroelasticity* was a poorly understood phenomenon until after World War I and was therefore responsible for some significant failures during it.

Aerofoil wing. A flat plate moving through a fluid such as air at a positive angle of attack will produce a certain amount of lift force. If the plate is curved in a certain way, however, it can be made to generate greater lift at the same angle of attack. A wing shaped in this fashion is known as an *aerofoil wing.* A typical aerofoil shape can be generated by mapping a circle using a conformal transformation in the form of a complex variable known as a Joukowski transform, named after the Russian mathematician and aerodynamicist who derived and published it in 1910.

Aileron and barrel rolls. There are fundamentally two techniques for rolling an aircraft. In an *aileron roll* the ailerons are used, often in conjunction with the rudder, to rotate the aircraft around its longitudinal axis while maintaining altitude. In a *barrel roll* the aircraft is guided around the outside of an imaginary barrel in the sky using a coordinated combination of aileron and elevator control. If performed correctly, the latter should be a 1g manoeuvre throughout.

Aspect ratio. The *aspect ratio* of a wing is its span divided by its mean chord, the latter being the average distance between the leading and trailing edges measured parallel to the normal air flow over the wing. A short-spanned wide wing would therefore have a *low aspect ratio* (LAR), whereas a long-spanned narrow wing would have a *high aspect ratio* (HAR). HAR wings produce significantly less induced drag (the drag produced by the process of generating lift) than LAR wings, but are much less manoeuvrable. A fighter aircraft would want manoeuvrability over reduced drag whereas a long-range glider would want the opposite, so a typical World War I fighter would have favoured an LAR design. Another reason for the shorter, wider wings generally seen during World War I would have been structural: it was difficult enough to stop the aircraft breaking up under aerodynamic loading as it was without adding the greater stresses that long wings naturally impose.

Calculus of variations. The *calculus of variations* is a branch of mathematics that can be used to seek out the path or surface for which a given function has a stationary value. Such values in the physical world are often associated with some sort of maximum or minimum. A typical example might be establishing the shortest distance between two points on some sort of surface. The history of the topic dates back to Isaac Newton in the late 1600s, and it was developed into a more mature field by Leonhard Euler and Joseph-Louis Lagrange during the following century.

Ceiling. An aircraft's absolute *ceiling* is the highest altitude at which it can sustain level flight given its weight and the atmospheric conditions.

Chord length. The *chord length* of an aerofoil is the length of an imaginary line that joins the trailing edge to the leading edge.

Controllable-pitch propeller. There are two generic types of propeller systems: fixed pitch and controllable pitch. Being able to control the pitch (or angle of attack into the relative air flow) of an aircraft propeller offers more finesse and greater flexibility in power and drag control.

Destructive testing. The practice of testing a structure or component to the point of failure is known as *destructive testing.* This is an expensive but essential element of any design process and helps define the limits within which the structure or component should be operated.

Fail safe. The *fail safe* ethos demands that the failure of certain components within a system leaves that system in a safe rather than dangerous condition. On an aircraft, a typical example would be an automatic feathering system on a propeller following an engine failure.

Finite-element analysis. *Finite-element analysis* is a technique that uses the computational power of modern-day computers to numerically solve partial differential equations that model physical phenomena and processes. A typical application would be defining the stresses and strains experienced by individual sections of some complex physical structure such as the metal frame of a tall building or a modern aircraft wing.

Form drag. *Form drag* is determined by the shape and size of a body moving through a fluid and arises because of the pressure difference created in front of and behind the object.

Horsepower. *Horsepower* is a unit of power measurement that was introduced with the advent of the steam engine in the 1700s. In SI units it equates to just under 750 watts: an approximate average of the power produced by a typical working horse.

Hysteresis. If a system of some description demonstrates *hysteresis*, then some property of that system lags behind the effect that is causing it. A typical example might be the magnetization of a piece of iron, where the change in the magnetism of the metal lags behind the fluctuations in the magnetic field responsible for inducing that magnetism.

Induced drag. *Induced drag* is drag that arises as a direct result of the process generating the lift on an aerofoil as it redirects the air in its vicinity. It is often referred to as *lift-induced drag.*

Lift-to-drag ratio. An aircraft encounters aerodynamic drag, D, as it moves through the air. It also generates lift from its wings, L. At any given speed one can calculate the *lift-to-drag ratio*, L/D, which is effectively a measure of the aircraft's aerodynamic efficiency.

Pitot. A *pitot* is a tube or probe used to measure flow in a fluid. In the context of an aircraft in flight, it senses the flow of air past the aircraft to provide an input into the air speed indicator. The term is often used in conjunction with the orifices that sense the static pressure surrounding the aircraft for input to the altimeter: the combination is known as the *pitot-static system.*

Stall. A wing *stall* occurs when the airflow across the top of the aerofoil detaches, causing a large decrease in lift. It generally happens when the angle of attack of a wing to the relative air flow becomes too high for the viscous forces pulling the air along the contour of the aerofoil to continue to do so. If the aircraft wings stall symmetrically, the nose of the aircraft drops, the angle of attack of the wings reduces, and normal flight can be regained if height allows. If one wing stalls before the other, a spin can result – something from which recovery is more challenging.

Standard operating procedures. All aircraft come with a set of instructions that define how to go about certain standard procedures: for example, how to fly a circuit and land. In this case it might define heights and speeds to fly, power settings during the various phases of the circuit, flare height, runway length required, and so forth. *Standard operating procedures* represent fundamental knowledge that every pilot should have at his or her fingertips to operate an aircraft safely.

Torque. A *torque* is the rotational equivalent of a linear force. It is a force that tends or attempts to cause a rotation of an object about a defined centre. Its value is calculated by multiplying the magnitude of a linear force by that force's perpendicular distance from a defined centre of rotation.

Whirling arm. A *whirling arm* is a device comprising a central pillar of sorts to which a mechanical arm is attached. Something such as an aerofoil section can be fixed to the end of the arm and the whole structure can be made to spin around, enabling study of the aerofoil's aerodynamic properties. Another way of thinking about the device is as the rotational equivalent of a linear wind tunnel.

Wing tip vortices. *Wing tip vortices* are a natural product of wing lift generation. The span-wise flow of air from wing root to wing tip coupled with air spilling around the wing from areas of higher to lower pressure, and the forward motion of the wing, form rotating vortices that trail from the rear of the wing tips.These vortices are often clearly visible when a large aircraft is landing in misty conditions.

Yaw, roll, and pitch. An aircraft has freedom about three mutually orthogonal axes. Rotation about the lateral axis is called *pitch*, rotation about the longitudinal axis is called *roll*, and rotation about the vertical axis is called *yaw*.

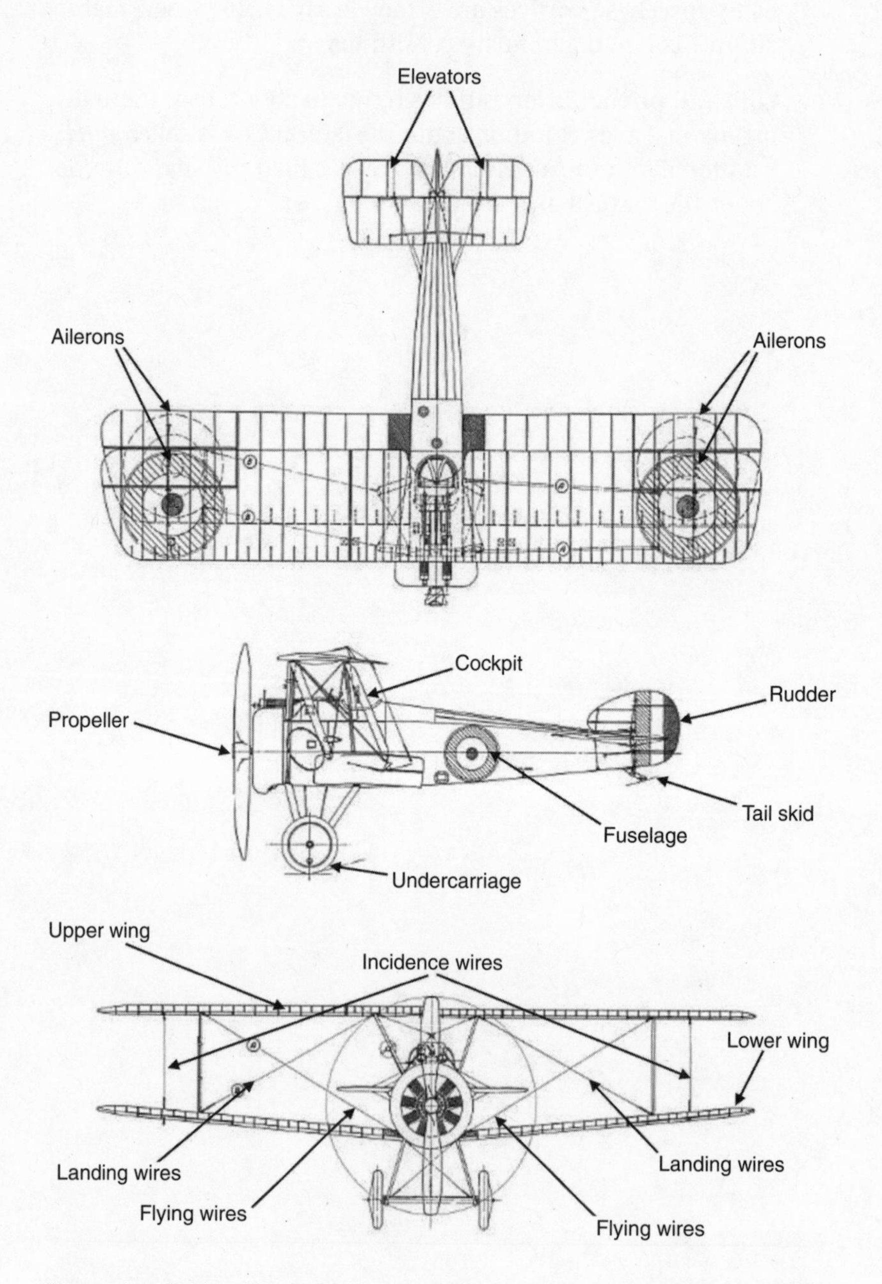

Figure A.4 Basic schematic of a Sopwith Camel (based on an original by William A. Wylam, 1947).

Figures

References

ARCHIVES

CAMBRIDGE

The papers and correspondence of Sir Geoffrey Ingram Taylor, GB 16 TAYLOR; and the papers and correspondence of Sir George Paget Thomson, GB 16 G.P. THOMSON. Both in the Trinity College Archives, University of Cambridge.

The papers and correspondence of Francis William Aston, GB 12 CUL F.W. Aston, Cambridge University Library.

OXFORD

The papers of F.A. Lindemann, Viscount Cherwell of Oxford, CSAC 80/4/81, in the Library of Nuffield College.

LONDON

The papers and correspondence of Alfred John Sutton Pippard, GB 98 B/PIPPARD, Archives and Corporate Record Unit, Imperial College.

The papers of John Turner MacGregor-Morris, JMM/2/3, Queen Mary University of London Archives.

The J.W. Dunne archive, DUNNE/A, Science Museum Library and Archives (held at Wroughton, Wiltshire).

MISCELLANEOUS

The personal flying logbooks of Sir Roderic Maxwell Hill, National Aerospace Library, Farnborough, Hampshire.

Ackroyd, J.A.D. 2002a. 'Sir George Cayley, the Father of Aeronautics. Part 1. The Invention of the Aeroplane'. *Notes and Records of the Royal Society* 56:167–81.

– 2002b. 'Sir George Cayley, the Father of Aeronautics. Part 2. Cayley's Aeroplanes'. *Notes and Records of the Royal Society* 56:333–48.

Ackroyd, J.A.D., and N. Riley. 2011. 'Hermann Glauert FRS, FRAeS (1892–1934)'. *Journal of Aeronautical History* 1:22–73.

Anderson, J.D. 1997. *A History of Aerodynamics and Its Impact on Flying Machines.* Cambridge: Cambridge University Press.

Anon. 1823. In *Flying in the Air*, ed. J. C. Robertson, Volume 1. London: Knight and Lacey.

– 1910. 'University and Educational Intelligence'. *Nature* 84:194.

– 1912. 'A.S.L. Flying School'. *Flight* IV:151.

– 1939. 'Charles Grey'. *Time Magazine*, Volume 34.

– 1957. 'Major R.H. Mayo'. *Flight* LXXI:294–5.

Argles, M. 1964. *South Kensington to Robbins: An Account of English Technical and Scientific Education Since 1851.* London: Longmans.

Aston, F.W., and G.P. Thomson. 1921. 'The Constitution of Lithium'. *Nature* 106:827–8.

Aubin, D., and C. Goldstein, eds. 2014. *The War of Guns and Mathematics: Mathematical Practices and Communities in France and Its Western Allies Around World War I.* Providence, RI: American Mathematical Society.

Bairstow, L. 1913. 'The Laws of Similitude'. *Flight* V:330–3.

Bairstow, L., B.M. Cave, and E.D. Lang. 1922. 'The Two-Dimensional Slow Motion of Viscous Fluids'. *Proceedings of the Royal Society of London* A 705:394–413.

– 1923. 'The Resistance of a Cylinder Moving in a Viscous Fluid'. *Philosophical Transactions of the Royal Society of London* A 614: 383–432.

Barrow-Green, J. 2014. In *Cambridge Mathematicians' Responses to the First World War*, ed. D. Aubin and C. Goldstein, 58–124. Providence, RI: American Mathematical Society.

– 2015. "'Anti-aircraft guns all day long": Karl Pearson and computing for the Ministry of Munitions'. *Revue d'Histoire des Mathématiques* 21:111–50.

Batchelor, G.K. 1958–71. *The Scientific Papers of Sir Geoffrey Ingram Taylor*, Volumes 1–4. Cambridge: Cambridge University Press.

Bellot, H.H. 1929. *University College London, 1826–1926.* London: University of London Press.

Benbow, T. 2011. *British Naval Aviation: The First 100 Years*. Farnham: Ashgate Publishing.
Berriman, A.E. 1911. 'Mathematics of the Cambered Plane'. *Flight* III:58–9.
- 1912a. *Aviation: An Introduction to the Elements of Flight*. London: Methuen & Company.
- 1912b. 'Parke's Dive'. *Flight* IV:787–9.
- 1914. 'Instruments Used in Experiments on Aeroplanes'. *Flight* VI:1097–8.
- 1915. *Motoring: An Introduction to the Car and the Art of Driving It*. London: Methuen & Company.
Bishop, R.E.D. 1987. 'Arthur Roderick Collar'. *Biographical Memoirs of Fellows of the Royal Society* 33:165–85.
Bloor, D. 2011. *The Enigma of the Aerofoil: Rival Theories in Aerodynamics, 1909–1930*. Chicago, IL: University of Chicago Press.
Boley, B.A. 2008. 'Castigliano, Carlo Alberto'. *Complete Dictionary of Scientific Biography* 3:117–19.
Bowyer, E. 1966. 'British Aircraft Industry – Men and Machines'. *Journal of the Royal Aeronautical Society* 70:280–4.
Boyd, T.J.M. 2011. 'One Hundred Years of G.H. Bryan's Stability in Aviation'. *Journal of Aeronautical History* 2011/4:97–115.
Brinkworth, B.J. 2014. 'On the Early History of Spinning and Spin Research in the UK. Part 1: The Period 1909–1929'. *Journal of Aeronautical History* 3:106–60.
Brodetsky, S. 1920. *A First Course in Nomography*. London: G. Bell & Sons.
- 1921a. 'Aeroplane Mathematics'. *Mathematical Gazette* 152:257–81.
- 1921b. *The Mechanical Principles of the Aeroplane*. London: J. & A. Churchill.
- 1921c. 'Aeronautics'. *Nature* 108:36–8.
Bryan, G.H. 1911. *Stability in Aviation: An Introduction to Dynamical Stability as Applied to the Motions of Aeroplanes*. London: Macmillan.
Bryan, G.H., and W.E. Williams. 1904. 'The Longitudinal Stability of Gliders'. *Proceedings of the Royal Society of London* 37:100–6.
Buchanan, R.A. 1985. 'The Rise of Scientific Engineering in Britain'. *British Journal for the History of Science* 18:218–33.
Busk, M. 1925. *E.T. Busk and H.A. Busk*. London: John Murray.
Castigliano, C.A. 1873. *Intorno ai Sistemi Elastici*. Turin: Bona.

Cave-Browne-Cave, B. 1922. 'The Calculations of the Periods and Damping Factors of Aeroplane Oscillations and a Comparison with Observations'. Technical Report 570, Advisory Committee for Aeronautics, 1918–1919.

Cave-Browne-Cave, F.E., and K. Pearson. 1902. 'On the Correlation between the Barometric Height at Stations on the Eastern Side of the Atlantic'. *Proceedings of the Royal Society of London* 70:465–70.

Cayley, A. 1870. 'On the Rational Transformation Between Two Spaces'. *Proceedings of the London Mathematical Society* 3:127–80.

Cayley, G. 1810. 'On Aerial Navigation: Parts 1, 2 and 3'. *Nicholson's Journal of Natural Philosophy, Chemistry, and the Arts* 25:81–7.

Chanute, O.A. 1894. *Progress in Flying Machines.* New York: American Engineer and Railroad Journal.

Charlton, T.M. 1974. 'Professor Bertram Hopkinson, C.M.G., M.A., B.Sc., F.R.S. (1874–1918)'. *Notes and Records of the Royal Society of London* 29:101–9.

Chitty, L. 1966. 'Contribution from L. Chitty'. *Journal of the Royal Aeronautical Society* 70:67–8.

Chitty, L., and R.V. Southwell. 1931. 'A Contribution to the Analysis of Primary Stresses in the Hull of a Rigid Airship'. *Journal of the Royal Aeronautical Society* 35:1103–36.

Clapeyron, P.B.E. 1858. 'Mémoire sur le Travail des Forces Élastiques dans un Corps Solide Élastique dÉformé par l'Action de Forces Extérieures'. *Comptes Rendus* 46:208–12.

Cokayne, G.E. 1902. *Complete Baronetage.* Exeter: W. Pollard & Co.

Collar, A.R. 1978a. 'The First Fifty Years of Aeroelasticity'. *Aerospace* 5:12–20.

– 1978b. 'Arthur Fage'. *Biographical Memoirs of Fellows of the Royal Society* 24:32–53.

Dawson, G.G. 1913. 'The Royal Society'. *The Times,* 8 May, p. 12.

De Hevesy, G.C. 1948. 'Francis William Aston, 1877–1945'. *Biographical Memoirs of Fellows of the Royal Society* 5:634–50.

Driver, H. 1997. *The Birth of Military Aviation: Britain, 1903–1914.* Woodbridge: Boydell & Brewer.

Dunne, J.W. 1927. *An Experiment with Time.* London: A. & C. Black.

Edgerton, D. 1991. *England and the Aeroplane: An Essay on a Militant and Technological Nation.* London: Macmillan.

Eiffel, G. 1910. 'La Résistance de l'Air et l'Aviation'. *Biographie Industrielle et Scientifique* 2:1–41.

Euler, L. 1744. *Methodus Inveniendi Lineas Curvas Maximi Minimive Proprietate Gaudentes, sive Solutio Problematis Isoperimetrici Lutissimo Sensu Accepti.* Geneva: Marcum-Michaelem Bousquet.

Fage, A., and L. Bairstow. 1920. 'Torsional Vibrations of the Tail of an Aeroplane. Part (II): Oscillations of the Tail Plane and Body of an Aeroplane in Flight'. Technical Report 276, Advisory Committee for Aeronautics, 1916–1917.

Fearon, P. 1969. 'The Formative Years of the British Aircraft Industry, 1913–1924'. *Business History Review* 4:476–95.

Flexner, H.H. 2014. 'The London Mechanics' Institution: Social and Cultural Foundations 1823–1830'. PhD Thesis, University College London.

Flood, R., A. Rice, and R. Wilson. 2011. *Mathematics in Victorian Britain.* Oxford: Oxford University Press.

Fort, A. 2003. *Prof: The Life of Frederick Lindemann.* London: Jonathan Cape.

Furinghetti, F. 2008. 'The Emergence of Women on the International Stage of Mathematics Education'. *ZDM: The International Journal of Mathematics Education* 40:529–43.

Gay, H. 2007. *The History of Imperial College London, 1907-2007: Higher Education and Research in Science, Technology, and Medicine.* London: Imperial College Press.

Gibbs-Smith, C.H. 1965. *Sir George Cayley's Aeronautics 1796–1855.* London: HMSO.

Glauert, H. 1919. 'The Investigation of the Spin of an Aeroplane'. Technical Report 618, Advisory Committee for Aeronautics, 1916–1917.

Greenhill, G. 1912. *The Dynamics of Mechanical Flight: Lectures Delivered at the Imperial College of Science and Technology, March, 1910 and 1911.* London: Constable & Company.

Grey, C.G. 1909. *Flying - The Why and the Wherefore.* London: Iliffe and Sons.

– 1912. 'Editorial'. *The Aeroplane* 2:123.

Hall, A.A. 1970. 'Sir William Farren, Hon. FRAeS, 1892–1970'. *Aeronautical Journal* 74:898.

Hashimoto, T. 1990. *Theory, Experiment, and Design Practice: The Formation of Aeronautical Research, 1909–1930.* PhD Thesis, Johns Hopkins University.

– 2000. In *The Wind Tunnel and the Emergence of Aeronautical Research in Britain,* ed. P. Galison and A. Roland. Dordrecht: Kluwer Academic.

Hill, M.J.M. 1884. 'On the Motion of Fluid, Part of Which Is Moving Rotationally and Part Irrotationally'. *Philosophical Transactions of the Royal Society of London* 175:363–409.

Hill, P. 1962. *To Know the Sky: The Life of Air Chief Marshal Sir Roderic Hill.* London: W. Kimber.

Hoff, N.J. 1946. 'A Short History of the Development of Airplane Structures'. *American Scientist* 34:370–88.

Howard, H.B. 1966. 'Aircraft Structures'. *Journal of the Royal Aeronautical Society* 70:54–66.

Hudson, H.P. 1912. 'On Binodes and Nodal Curves'. *Proceedings of the Fifth International Congress of Mathematicians (Cambridge, 1912)*, Volume II, 118–121. Cambridge: Cambridge University Press.

- 1920a. 'Incidence Wires'. *Aeronautical Journal* 24:505–16.

- 1920b. 'The Strength of Laterally Loaded Struts'. *The Aeroplane* 18:1178–80.

- 1927. *Cremona Transformations: In Plane and Space.* Cambridge: Cambridge University Press.

Hume-Rothery, J.H. 1912. 'The X Constant'. *Flight* IV:841.

- 1913a. 'Negative Wing Tips and Lateral Stability'. *Flight* V:64–5.

- 1913b. 'The Vol Piqué'. *Flight* V:1047–9.

- 1913c. 'The Vol Piqué'. *Flight* V:1075–6.

Hunsaker, J.C. 1916. 'Dynamical Stability of Aeroplanes'. *Proceedings of the National Academy of Sciences* 2:278–83.

Jakab, P.L. 1999. 'Wood to Metal: The Structural Origins of the Modern Airplane'. *Journal of Aircraft* 36:914–18.

Jones, C.G. 2009. *Femininity, Mathematics and Science, 1880–1914.* London: Palgrave Macmillan.

- 2010. In *Femininity and Mathematics at Cambridge Circa 1900*, ed. J. Spence, S. Aiston, and M. Meikle. New York: Routledge.

Jones, R.V. 1987. 'Lindemann Beyond the Laboratory'. *Notes and Records of the Royal Society* 41:191–210.

Kimber, E., and R. Johnson. 1771. *The Baronetage of England.* London: G. Woodfall.

Kinney, J. 2017. *Reinventing the Propeller: Aeronautical Speciality and the Triumph of the Modern Airplane.* Cambridge: Cambridge University Press.

Lanchester, F.W. 1907. *Aerodynamics: Constituting the First Volume of a Complete Work on Aerial Flight.* New York: D. Van Nostrand.

- 1908. *Aerodonetics.* London: Constable & Company.

- 1914. 'The Flying Machine from an Engineering Standpoint'. *Flight* VI:523–6.

- 1916. 'Development of the Military Aeroplane: The Question of Size'. *Engineering* 101:212–14.

Langley, S.P. 1891. 'Experiments in Aerodynamics'. *Smithsonian Contributions to Knowledge*, Publication 801, Volume 27, Number 1. Washington, DC: Smithsonian Institution.

Laurence, J. 1938. *Murder in the Stratosphere.* London: Sampson Low.

Lindemann, F.A. 1917a. 'Note on R. & M. No. 310 - "Note on a Rate of Climb Indicator for Use on Aeroplanes"'. Report T.910, March, Royal Aircraft Factory H-Department. Lindemann Archive.

- 1917b. 'Note on Recording Accelerometer for Measuring Stresses on Aeroplanes'. Report 621, 15 June, Royal Aircraft Factory H-Department. Lindemann Archive.

Lindemann, F.A., and F.W. Aston. 1919. 'The Possibility of Separating Isotopes'. *London, Edinburgh, and Dublin Philosophical Magazine and Journal of Science* 37:523–34.

Lindemann, F.A., H. Glauert, and R.G. Harris. 1918. 'The Experimental and Mathematical Investigation of Spinning'. Technical Report 411, Advisory Committee for Aeronautics, 1916–1917.

Lovelace, A. 1843. 'Notes on L. Menabrea's Sketch of the Analytical Engine Invented by Charles Babbage, Esq'. *Taylor's Scientific Memoirs* 3:691–731.

Lucas, K. 1914. 'Report on the Errors of Compasses on Aeroplanes'. Report 251, undated, Royal Aircraft Factory A.E.Ph. Lucas Archive.

- 1915a. 'Letter from Lucas to an Unknown Recipient'. Letter, 25 March. Lucas Archive.

- 1915b. 'Movement Shown by Longitudinal Level in an Aeroplane During Pitch Oscillations'. Report 672, Royal Aircraft Factory H-Department. Lucas Archive.

Macaulay, W.H. 1913. *The Laws of Thermodynamics.* Cambridge: Cambridge University Press.

Marsden, B. 1992. 'Engineering Science in Glasgow: Economy, Efficiency and Measurement as Prime Movers in the Differentiation of an Academic Discipline'. *British Journal for the History of Science* 25:319–46.

McComas, A.J. 2011. *Galvani's Spark - The Story of the Nerve Impulse.* Oxford: Oxford University Press.

McKinnon Wood, R. 1960. 'Memories of Chudleigh Mess'. *New Scientist* 7:1532–4.

Menabrea, L.F. 1842. 'Notions sur la Machine Analytique de M. Charles Babbage'. *Bibliothèque Universelle de Genève* 3:352–76.

O'Gorman, M. 1916a. 'Letter from O'Gorman to the Secretary of the Advisory Committee for Aeronautics'. Letter, 9 October. Lucas Archive.

– 1916b. 'Improvements in Magnetic Compasses'. Report 537, Royal Aircraft Factory H-Department. Lucas Archive.

Pemberton Billing, N. 1916. *Air War: How to Wage It.* Chatham: Gale & Polden.

Perry, J., and W.E. Ayrton. 1886. 'On Struts'. *The Engineer* 62:464–5, 513–14.

Pinsent, D.H. 1921. 'Exploration of the Slipstream Velocity in a Pusher Machine'. Technical Report 444, Advisory Committee for Aeronautics, 1917–1918.

Pinsent, D.H., and H.A. Renwick. 1921. 'The Variation of Engine Power with Height'. Technical Report 462, Advisory Committee for Aeronautics, 1917–1918.

Pippard, A.J.S. 1966. 'Admiralty Air Department 1915'. *Journal of the Royal Aeronautical Society* 70:69–70.

Pippard, A.J.S., and J.L. Pritchard. 1918. *Handbook of Strength Calculations.* London: Ministry of Munitions Technical Department (Aircraft Production).

– 1919. *Aeroplane Structures.* London: Longmans, Green Co.

Porter, T.M. 2010. *Karl Pearson: The Scientific Life in a Statistical Age.* Princeton, NJ: Princeton University Press.

Pudney, J. 1948. *Laboratory of the Air.* London: HMSO.

Pugsley, A.G. 1961. 'Robert Frazer'. *Biographical Memoirs of Fellows of the Royal Society* 7:78–80.

Rayleigh, Lord. 1876. 'On the Resistance of Fluids'. *London, Edinburgh and Dublin Philosophical Magazine and Journal of Science* 44:430–41.

– 1877. *The Theory of Sound* (two volumes). London: Macmillan.

– 1891. 'Experiments in Aerodynamics'. *Nature* 45:108–9.

– 1900. 'On the Mechanical Principles of Flight'. *Memoirs and Proceedings of the Manchester Literary & Philosophical Society* 5:1–26.

– 1915. Technical Report of the Advisory Committee for Aeronautics, 1913–1914, p. 21.

Reese, P. 2014. *The Men Who Gave Us Wings: Britain and the Aeroplane 1796–1914.* Barnsley: Pen and Sword.

Reynolds, O. 1883. 'An Experimental Investigation of the Circumstances which Determine whether the Motion of Water Shall Be Direct or Sinuous, and of the Law of Resistance in Parallel Channels'. *Philosophical Transactions of the Royal Society of London* A 174:935–82.

Robertson, J.C. 1824. 'Editor's Preface to Volume 1 of *Mechanic's Magazine*'. London: Knight and Lacey.

Rood, G. 2011. 'Edward Teshmaker Busk'. History and Learning Briefing 10, Farnborough Air Sciences Trust.

Routh, E.J. 1877a. *An Elementary Treatise on the Dynamics of a System of Rigid Bodies.* London: Macmillan.

– 1877b. *A Treatise on the Stability of a Given State of Motion.* London: Macmillan.

Royle, A.P. 2017. 'The Impact of the Women of the Technical Section of the Admiralty Air Department on the Structural Integrity of Aircraft during World War One'. *Historia Mathematica* 44:342–66.

Segre, C. 1897. 'Sulla Scomposizione dei Punti Singolari delle Superficie Algebriche'. *Annali di Matematica Pura ed Applicata* 25:1–54.

Shute, N. 1961. *Stephen Morris.* London: Vintage Books.

Skempton, A.W. 1970. 'Alfred John Sutton Pippard'. *Biographical Memoirs of Fellows of the Royal Society* 16:463–78.

Smith, M. 1953. 'C.G.G.'. *Flight and Aircraft Engineer* 64:803.

Southwell, R.V. 1912. 'The Strength of Struts'. *Engineering* 94:249–50.

– 1913. 'On the General Theory of Elastic Stability'. *Philosophical Transactions of the Royal Society of London* A 213:187–244.

– 1915a. 'On the Collapse of Tubes by External Pressure. Part I'. *London, Edinburgh and Dublin Philosophical Magazine and Journal of Science (Series 6)* 70:687–98.

– 1915b. 'On the Collapse of Tubes by External Pressure. Part II'. *London, Edinburgh and Dublin Philosophical Magazine and Journal of Science (Series 6)* 153:502–11.

– 1915c. 'On the Collapse of Tubes by External Pressure. Part III'. *London, Edinburgh and Dublin Philosophical Magazine and Journal of Science (Series 6)* 169:67–77.

Southwell, R.V., and L. Chitty. 1930. 'On the Problem of Hydrodynamic Stability. I. Uniform Shearing Motion in a Viscous Fluid'. *Philosophical Transactions of the Royal Society of London* A 229: 205–53.

Spooner, S. 1913. 'Editorial'. *Flight* V:54.

Squier, G.O. 1909. 'Present State of Military Aeronautics'. *Flight* I:121–3.

Stephens, A.V. 1966. 'Some British Contributions to Aerodynamics'. *Journal of the Royal Aeronautical Society* 70:71–8.

Stoney, B. 2000. *Twentieth Century Maverick – The Life of Noel Pemberton Billing.* New Romney: Bank House Books.

Taylor, G.I. 1966. 'When Aeronautical Science Was Young'. *Journal of the Royal Aeronautical Society* 70:108–13.

Temple, G., A. Fage, J.L. Nayler, and E.F. Relf. 1965. 'Leonard Bairstow'. *Biographical Memoirs of Fellows of the Royal Society* 11:22–40.

Thompson, F.M.L. 1990. *University of London and the World of Learning, 1836–1986.* London: Bloomsbury.

Thomson, G.P. 1915. 'Calculations on the Spinning of an Aeroplane'. Technical Report 211, Advisory Committee for Aeronautics 1915–1916.

- 1919. *Applied Aerodynamics.* London: Hodder and Stoughton.

- 1946. 'Dr. Francis William Aston, F.R.S.'. *Nature* 157:290–2.

Thomson, G.P., and A.A. Hall. 1971. 'William Scott Farren, 1892–1970'. *Biographical Memoirs of Fellows of the Royal Society* 17:214–41.

Thurston, A.P. 1949. 'Reminiscences of Early Aviation'. *Transactions of the Newcomen Society* 1:1–6.

Tournès, D. 2016. 'Maurice d'Ocagne's Papers on Nomography'. Paper presented at *CIRMATH – Circulating Mathematics via Journals: The Rise of Internationalization 1850–1920, June 20–23, 2016, Institut Mittag-Leffler.*

Turner, P.K. 1912. 'Second Thoughts of an Idle Designer'. *The Aeroplane* 2:282.

von Helmholtz, H. 1891. 'On a Theorem Relative to Movements that Are Geometrically Similar in Fluid Bodies, Together with an Application to the Problem of Steering Balloons'. In *Mechanics of the Earth's Atmosphere: A Collection of Translations*, ed. C. Abbe, 67–77. Smithsonian Miscellaneous Collections, Volume 843. Washington, DC: Smithsonian Institution.

von Wright, G.H. 1990. *A Portrait of Wittgenstein as a Young Man.* Oxford: Blackwell.

Walker, P.B. 1974. *Early Aviation at Farnborough: The First Aeroplanes.* London: Macdonald.

Warwick, A. 2003. *Masters of Theory: Cambridge and the Rise of Mathematical Physics*. Chicago, IL: University of Chicago Press.

Wilson, E.B. 1918. 'The Mathematics of Aerodynamics'. *American Mathematical Monthly* 25:292–7.

Wittgenstein, L.J.J. 1922. *Tractatus Logico-Philosophicus*. London: Trubner & Co.

Index